AF393684

Jörg Ettrich

Fluid Flow and Heat Transfer in Cellular Solids

Schriftenreihe
des Instituts für Angewandte Materialien
Band 39

Karlsruher Institut für Technologie (KIT)
Institut für Angewandte Materialien (IAM)

Eine Übersicht aller bisher in dieser Schriftenreihe erschienenen
Bände finden Sie am Ende des Buches.

Fluid Flow and Heat Transfer
in Cellular Solids

by
Jörg Ettrich

Dissertation, Karlsruher Institut für Technologie (KIT)
Fakultät für Maschinenbau
Tag der mündlichen Prüfung: 04. Juli 2014

Impressum

Scientific
Publishing

Karlsruher Institut für Technologie (KIT)
KIT Scientific Publishing
Straße am Forum 2
D-76131 Karlsruhe

KIT Scientific Publishing is a registered trademark of Karlsruhe
Institute of Technology. Reprint using the book cover is not allowed.

www.ksp.kit.edu

Print on Demand 2014

ISSN 2192-9963
ISBN 978-3-7315-0241-8
DOI 10.5445/KSP/1000042041

For my beloved wife Melanie
For Samira and Cheyenne

For my beloved parents
Hans and Ute

And to the memory of
Emil and Elisabeth Jost

Oh me, Oh life
Of the question of these recurring,
Of the endless train of the faithless,
Of cities fill'd with the foolish,
What good admit these,
Oh me, Oh life

That thou art here – that life exists and identity,
that the powerful play goes on, and you may contribute a verse.

– Walt Whitman (1819-1892) –

Kurzdarstellung

Metallschäume sind funktionale Werkstoffe, deren zelluläre Strukturen natürlichen Materialien wie z.B. Knochen oder Pflanzen nachempfunden sind. Aufgrund ihrer hohen Steifigkeit bei geringer Masse, werden geschlossenporige Metallschäume bereits seit Längerem als Konstruktionselemente im Leichtbau, als Energieabsorptionselemente, als Schwingungsdämpfer oder zur Schalldämmung eingesetzt. Dagegen sind offenporige Metallschäume wegen ihrer durchströmbaren Struktur, ihrer sehr großen benetzbaren Oberfläche und ihres großen Verhältnis von Oberfläche zu Volumen für eine Reihe weiterer Anwendungen von großem Interesse, z.B. als Wärmeübertragungselemente oder Katalysatoren in der Verfahrenstechnik.

Um die für eine Anwendung notwendigen charakteristischen, thermischen und strömungsmechanischen Eigenschaften zu bestimmen und einen systematischen praxisgerechten Einsatz mit Hilfe von Korrelationen sowie Modellansätzen zu ermöglichen, werden im Rahmen dieser Arbeit zum einen experimentelle Untersuchungen durchgeführt und numerische Methoden für die Simulation der Strömung und Wärmeübertragung entwickelt. Die experimentellen Untersuchungen tragen einerseits zur Erweiterung der heute verfügbaren Datenbasis, speziell für Reynoldszahlen im Bereich von $800 \lesssim Re \lesssim 35'000$ bei, und dienen andererseits als Validierungsgrundlage für Simulationsstudien. Die durchgeführten Messungen umfassen dabei neben der Charakterisierung der Schaumstruktur anhand von Porendurchmesser, Stegdurchmesser, relativer Dichte und Porosität auch die Vermessung der Druckverluste und des Wärmeübertragungsverhaltens für unterschiedliche durchströmte Längen für den o.g. Reynoldszahlenbereich.

Im Vergleich zu den bereits etablierten und bekannten numerischen Methoden der Strömungsmechanik und Wärmeübertragung ist die Phasenfeldmethode ein recht junges und dennoch sehr mächtiges Werkzeug. Gerade im Hinblick auf die heutigen Anforderungen interdisziplinärer Aufgabenstellungen und die gekoppelte Multiphysiksimulation, birgt sie enormes Potential für die Entwicklung eines nachhaltigen Simulationsverfahrens. In dieser Arbeit werden ein Lattice-Boltzmann Strömungslöser und ein Lösungsverfahren für die Berechnung der Wärmeübertragung im Kontext einer Phasenfeldmethode entwickelt, getestet und angewendet. Von besonderer Bedeutung ist hierbei die numerische Modellierung im Bereich der diffusen Grenzflächen. Während für den Strömungslöser bekannte Konzepte erfolgreich angewendet werden können, wird für die Energiegleichung ein neuer Ansatz für die Formulierung der dreidimensionalen instationären Wärmeübertragung mit Hilfe einer tensoriellen Beschreibung der Mobilitäten und einem angepassten Divergenzoperator vorgestellt. Abschließend wird die Anwendbarkeit am Beispiel offenporiger Metallschäume demonstriert und ein qualitativer und quantitativer Vergleich zu experimentell ermittelten Kenngrößen gegeben.

Abstract

Metal foams are functional engineered materials, motivated by the structures of natural materials, e.g. bones or plants. Due to their high stiffness and low mass, closed-pore metal foams have been underway for some time as construction elements in lightweight construction, energy-absorbing elements, as vibration dampers or used for sound insulation. In contrast, open-cell metal foams, due to their permeable structure, its very large wettable surface and their large ratio of surface area to volume, are of great interest for a number of other applications, e.g. as heat transfer elements or catalysts in chemical engineering.

To determine the necessary characteristic, thermal and fluid mechanical properties for an application, and to allow a systematic practical use by means of correlations and modeling approaches, in the present work we perform experimental investigations and develop numerical methods for the simulation of flow and heat transfer. The experimental investigations on one-hand contribute to enlargement of the currently available data base, especially for Reynolds numbers in the range of $800 \lesssim \mathrm{Re} \lesssim 35'000$, and on the other hand, serve as validation basis for simulation studies. Besides to the characterization of the foam structure based on pore-diameter, edge-diameter, specific gravity and porosity, the measurements comprise the acquisition of pressure loss and the heat transfer performance for different lengths and the above-mentioned range of Reynolds numbers.

Compared to the already established and well-known numerical methods of fluid mechanics and heat transfer, the phase-field method is a relatively young, yet very powerful tool. Especially in view of the current requirements of interdisciplinary tasks and coupled multi-physics simulations, it harbours an enormous potential for the development of a sustainable simulation method. In the present work, a lattice Boltzmann and a heat transfer solver are developed, tested and applied in the context of a phase-field method. Of particular importance is the numerical modelling in the area of the diffuse interface. While we can successfully apply known and established methods for the flow solver, we present a new approach for the three-dimensional transient heat transfer, by means of a tensorial mobility approach and a revised divergence operator. Finally, the applicability is demonstrated using the example of open-cell metal foams and given a qualitative and quantitative comparison with experimentally determined parameters.

Acknowledgement

First of all, I would like to give my sincerest thanks to Prof. Dr. rer. nat. Britta Nestler and Prof. Dr.-Ing. Eckhard Martens, for putting their trust in me and facilitating this work.

Support provided by the Federal Ministry of Education and Research (BMBF) through the FHProfUnt funding project SimFoam (FKZ 17029X10) is greatly acknowledged.

For the endless and fruitful discussions on physics, for untiring support in maths (»*oh yes, tensors,. . . what was this all about?*«), for generous support in software and programming issues (»*so what the hell does* `malloc` *and what's about all these pointers, and hey, how to use* `git`*?*«), for the energetically support in experimental work and for any other support I would like to express my deepest gratitude to A. August, M. Ben-Said, M. Berghoff, J. Buch, A. Choudhury, Ch. Gaber, T. Hensgen, J. Hötzer, C. Hertweck-Maurer, M. Jainta, U. Kappler, A. Kneer, A. Lichtenberger, Ch. Mennerich, Ch. Ratz, M. Reichhart, D. Schneider, E. Schoof, E. Schulz, S. Schwab, M. Selzer, O. Thümmes, O. Tschukin, A. Vondrous and. . . if someone is missing on this list, please be lenient, I will keep all of you in mind for the great time and fellowship at the Institute.

Finally, I would like to express my most cordial and warmest gratitude to my beloved parents Hans and Ute, my beloved wife Melanie and my two wonderful and patient daughters Samira and Cheyenne (»*It did take some time, but now we're finally done*«).

Jörg Ettrich
Karlsruhe, Mai 2014

Contents

List of Tables

Listings

Nomenclature

Characteristic numbers

Gr	Grashof number
Hg	Hagen number
Kn	Knudsen number
Ma	Mach number
Pe	Peclet number
Pr	Prandtl number
Re	Reynolds number

Greek symbols

α	phase identifier
β	phase identifier
γ	surface tension
δ_x	discrete spatial spacing
δ_t	discrete temporal spacing
Δ	filter size
ϵ	perturbation parameter
ϵ	model parameter related to interface width
ζ	non-dimensional axial coordinate
ϑ	temperature
θ	potential
λ	finite interface width
μ	dynamic viscosity
ν	kinematic viscosity
$\boldsymbol{\xi}$	molecular velocity
$\boldsymbol{\Pi}, \Pi_{ij}$	non-equilibirum momentum tensor
ε	model parameter related to the finite interface width
ρ, ρ^*	density, specific gravity
ς	bulk viscosity
τ	relaxation time
τ_k	kinetic coefficient
ω	collision frequency
ϕ	order parameter
χ	volume fraction

ψ	porosity
Λ	Lagrange multiplier
Ω	Domain of interest

Indices

0	reference/characteristic value
c	concentration
coll	collision
e	energy
eq	equilibrium
f, fluid	fluid/liquid phase
fi	fluid inlet
fo	fluid outlet
hydr	hydraulic
lb	lattice units
p	physical units
obst	multi-obstacle potential
ref	reference
s, solid	solid phase
t	turbulent
therm	thermal
W	wall
well	multi-well potential
$*$	effective values (Smagorinsky subgrid scale model)
i, j, k, l	spatial directions
n, t, s	co-ordinate system aligned with the diffuse interface
α, β	identifiers to distinguish phases

Roman symbols

$a(\phi, \nabla\phi)$	surface energy function
a	thermal diffusivity
A	surface area
c	concentration
c_P	specific heat at constant pressure
c_s	speed of sound (in lattice units)
c_V	specific heat at constant volume
C_e	electric cpacitance
C_V	heat capacity
d	number of spatial dimensions
D	coefficient of diffusion
e_i	lattice vector for direction i
e	internal energy

f	free energy
f_i	particle distribution function along direction i
$\mathbf{F}$	Force
g	gravity
G	Gibbs-Simplex
$h, h(\phi)$	interpolation function
I	thermal inertia
$\boldsymbol{j}$	flux vector
k	thermal conductivity
$\mathbf{K}$	thermal conductivity tensor
l	length
$\mathcal{L}$	linear term of the segmented tensorial formulation
m	mass
M	mobility coefficient
$\mathcal{M}$	multiple relaxation time transformation matrix
$\mathcal{M}_{\mathcal{R}}$	modified divergence operator matrix multiplier
$n, \boldsymbol{n}$	normal coordinate, interface normal
N	number of particles
p	pressure
q	number of discrete lattice velocities
$\boldsymbol{q}$	heat flux vector
$\boldsymbol{q}_{\alpha\beta}$	generalised interface normals
Q	electric charge
$\dot{Q}$	rate of heat flow
$\mathbf{Q}$	projection onto interface normal
$\mathcal{Q}$	quadratic term of the segmented tensorial formulation
r	radius
R	resistance
$\mathcal{R}$	ratio of interpolation functions in modified divergence operator
R_{therm}	thermal resistance
s	entropy density
$\boldsymbol{s}$	vector of relaxation parameters for the multiple relaxation time model
S	entropy
$\boldsymbol{\mathcal{S}}, \mathcal{S}_{ij}$	strain rate tensor
t	time
T	temperature
$\mathbf{T}, \mathsf{T}_{ij}$	deviatoric stress tensor
$\boldsymbol{u}, u$	velocity
V	volume
$\dot{V}$	volume flow
w_i	weights of the equilibirum distribution function
$w(\phi)$	potential function
W_{hydr}	hydraulic power

x vector of spatial coordinates

Operators

$\overline{\cdot}$ mean
$\langle\cdot\rangle_{(A)}$ surface averaged value
$\langle\cdot\rangle_{(I)}$ mass averaged value

Acronyms

BGK   Bhatnagar-Gross-Krook approximation of the collision operator
CAD   Computer Aided Design
CFD   Computational Fluid Dynamics
CFL   Courant Friedrichs Lewy
D2Q9   two-dimensional LB model with nine velocities
D3Q19   three-dimensional LB model with nineteen velocities
DFG   Deutsche Forschungsgemeinschaft
HPC   high performance computing
HTLBE   Hybrid Thermal Lattice Boltzmann Equation
LB   lattice Boltzmann
LBM   lattice Boltzmann method
LES   large eddy simulation
LGA   lattice gas automaton
MPI   Message Passing Interface
MRT   multiple relaxation time model
NS   Navier Stokes
PACE3D   Parallel Algorithm for Crystal Evolution in 3D
ppi   pores per inch
RANS   Reynolds Averaged Navier Stokes
REV   representative elementary volume
SGS   subgrid scale
SRT   single relaxation time model
STL   Standard Tessellation Language

Mathematical typesetting

Value / operation	Tensorial notation	Index notation
Scalars		$a = \alpha \sqrt{\gamma c}$
Vectors	$\mathbf{u}$	u_i
Matrix	M	M_{ij}
Tensors	$\mathbf{T}$	$T_{\alpha\beta}$
Scalar product	$\lambda = \mathbf{a} \cdot \mathbf{b} = \langle \mathbf{a}, \mathbf{b} \rangle$	$\lambda = a_i b_i$
Cross product	$\mathbf{c} = \mathbf{a} \times \mathbf{b}$	$c_\alpha = \varepsilon_{\alpha\beta\gamma} a_\beta b_\gamma$ (†)
Gradient	$\mathbf{a} = \nabla\lambda$	$a_\alpha = \partial_\alpha \lambda$
Divergence of vector	$\lambda = \nabla \cdot \mathbf{a} = \langle \nabla, \mathbf{a} \rangle$	$\lambda = \partial_\alpha a_\alpha$
Divergence of 2D tensor	$\mathbf{a} = \nabla \cdot \mathbf{A} = \langle \nabla, \mathbf{A} \rangle$	$a_\alpha = \partial_\beta A_{\alpha\beta}$
Tensor contraction	$\lambda = \mathbf{A} : \mathbf{B}$	$\lambda = A_{\alpha\beta} B_{\alpha\beta}$
Dyadic vector product	$\mathbf{A} = \mathbf{a} \otimes \mathbf{b}$	$A_{\alpha\beta} = a_\alpha b_\beta$

†: $\varepsilon_{\alpha\beta\gamma}$ is the Levi-Civita symbol

Chapter 1

Introduction

Along with the rise of today's issues such as energy-efficiency and thermal balance of machinery, the interest on cellular solids in engineering applications has increased. During the last decades metal foam, which belongs to the class of porous media, was established in engineering applications. At first, mostly closed cell metal foams were used for structural components because of their unique mechnical features. Subsequently open cell metal foams were identified for the use in chemical or heat transfer applications, accounting for their huge surface to volume ratio and their penetrable structure [7, 113, 189]. Referring to [86] a cellular solid is made up of an interconnected network of solid struts or plates which form the edges and faces of cells. These cellular solids – to which open cell metal foams adhere – are characterized by the same physical, mechanical and thermal properties as pure solids, whereas the range and combination of these properties distinguish their inimitability and opens their application to a variety of engineering applications and innovative products.

Although research on porous media has been conducted since the mid 18th century, where Darcy deduced the basic equations and characteristics of flow in porous media, fluid flow and thermal physics in cellular solids still cannot be verified in a general fashion. Even though the physical and engineering sciences are aware of this topic since more than about 150 years, it has not lost its actuality, and there are still open issues to be clarified. At this point, experimental as well as numerical investigations can help to gain a deeper insight and understanding.

While the early publications are mainly focused on the mechanical properties, a rewiev on the last decades literature illustrates the emerging importance of the fluidic and thermal application of open cell metal foams. In order to lead the basic flow properties into the design process of related applications and products, experimental and computational efforts are mandatory. From 1999 until 2004 the priority program 1075 on *Cellular Metallic Materials*[1], concerned with the properties, processing parameters, optimisation and understanding of the new materials, was founded by the Deutsche Forschungsgemeinschaft (DFG). Apart from that, numerous experimental and numerical efforts underline the importance and actuality of that topic to this day. A recent review on heat transfer applications of matrix and foam materials is given in [94] and the performance of finned heat exchangers and open cell metal foam heat exchangers is given in [183]. In [132–135, 222] experimental pressure loss and heat transfer measurements are conducted for a large number of aluminium and copper foams, whereas [149] investigated the pressure loss characteristics in nickel foams. Ceramic foam structures are in the focus of [48, 90] for applications in catalytic and chemical applications. For all publications mentioned, the range of Reynolds numbers is limited to a small to medium range. Thus, the range of medium to high Reynolds numbers, which is in particular interesting for engineering applications, will be covered by the present work.

There are only few publications covering the numerical treatment of the fluid flow and heat transfer in cellular solids on a detailed high resolution pore scale level, cf. [12, 19, 38, 46, 114, 123] to name only a few, and which are discussed later in the text and which is the motivation for the development of a straightforward, seamless and comfortable method.

1.1 Scope of this work

As a representative for the group of cellular solids, the focus of the present work is laid on open cell metal foams. The objective of the thesis is twofold: at first, the experimental investigations contribute to the available database of hydraulic and thermal properties of open cell metal foams, whereas the modelling efforts contribute to the development of appropriate numerical methods in the context of the diffuse interface phase field approach.

For the experiments, the measurement range is chosen in order to cover a moderate to high Reynolds number flow regime, since there is only few data available. From this, we claim to establish correlations for pressure loss and thermal performance. The numerical work will represent the capabilities of the established methods and algorithms for

---

[1]`http://gepris.dfg.de/gepris/projekt/5469604` (accessed: 20/4/2014)

their use on complex microstructures and cellular solids, by means of focusing exemplarily on open cell metal foams, as a predecessor of a more extensive parameter study in comparison to experimental data.

In view of more complex multidisciplinary and multiphysics applications e.g. including phase transition, electro chemistry or solute transport, as well as more complex convoluted microstructures, we compile our work in the context of a phase field method [154]. The phase field method emerged from simulations within the scope of solidification, crystal-growth or phase-transition problems. In this contribution, we will utilize the method for the creation and mapping of complex pore-scale structures. Thus, no phase-field dynamics is present, i.e. the phase field is only used to distinguish between different phases (fluid I solid) which do not evolve in time. One can easily think of applications, topologies and configurations which can hardly be mastered by body fitted, interface tracking methods, if ever. For the latter situations, the presented approach promise to be a comfortable, feasible and high potential method.

For the solution of the fluid flow, we will employ the lattice Boltzmann method, which has evolved as a powerful and valuable tool, on eye-level with classical computational fluid dynamics methods, for computational studies of incompressible and quasi-incompressible flow regimes. However, there are only a few publications on lattice Boltzmann models for fluid flow in the context of phase field throughout the last decade. Whereas [148, 175, 176] are concerned with the simulation of dendritic solidification including the effects of liquid motion, [67, 155] aim at the simulation of fluid flow in complex microstructures unlike the *classical* field of application of a phase-field method.

Regarding heat transfer simulations, the evolution equation of temperature is solved within the framework of the phase-field method. In the course of this work, the numerical treatment is subject of the development of a new segmented tensorial formulation, which allows for the stable and accurate solution of three dimensional transient temperature field.

The solution of the Navier-Stokes equations is accompanied by several numerical difficulties, the treatment of nonlinear convective terms or the solution of the Poisson equation to evaluate the pressure [161]. In lattice Boltzmann methods, the pressure is obtained by the equation of state and the nonlinear convective terms reduce to an advection type problem [219]. Furthermore, the LBM is completely explicit, thus simple to implement. Each timestep is divided into a collision and a streaming step, which are local and nearest neighbour operations and which are hence predestined for parallelisation [95, 194, 215].

1.2 Outline of the thesis

The thesis is organized as follows: In chap. 2 we describe the main properties of open cell metal foams, their characterisation and a new method which enables the straightforward, feasible and realisitc modelling of foam structures, followed by the experimental setup as well as the results of fluid dynamic and thermal measurements, discussed in chap. 3. The presentation of the numerical approach starts in chap. 4, where a brief overview on the idea of a diffuse interface phase field method is given. The integration of established numerical approaches as well as the development of new methods for diffuse interface fluid dynamics and heat transfer together with results of their validation, is provided in chap. 5 and chap. 6, respectively. Results of the modelling of real foam structures, in terms of coupled diffuse interface fluid flow and heat transfer approach, are evaluated and compared to measurements of real foam structures in chap. 7. Finally an outlook on potential improvements, optimisations and future developments is given in chap. 8, and the thesis is closed by a summary and conclusion in chap. 9.

Chapter 2

Metal foam

This rather young group of materials is the focus of attention of academics and engineers since the early nineties. Similarly to bionic structures or polymer foams, they have extremely low densities and high specific surface areas. Utilising metals as base materials results in excellent mechanical, electrical, thermal and acoustic properties, whereas specific properties like thermal expansion coefficient, electric conductivity or melting temperature remain unchanged compared to the raw material [7].

Metal foams can be made from almost all base materials, whereas each one shows up its individual and specific foam structure. Due to their low melting temperatures, aluminium, magnesia, tin and copper are

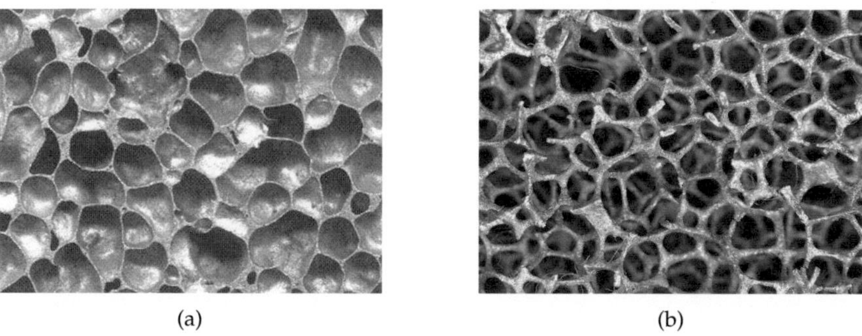

(a)   (b)

Figure 2.1: Close up of (a) closed cell foam[1] and (b) open cell foam (pore diameter of approximately 5 mm each).

[1]This image is taken from: `http://commons.wikimedia.org/wiki/File:Closed_cell_metal_foam_with_large_cell_size.JPG`, available under the Creative Commons Attribution/Share-Alike Licence. (accessed: 10/3/2014)

appropriate for a reasonable and cost-efficient production [100]. The cell structure of metal foam is of statistical nature, where the cells and the pore sizes are more or less statistically scattered, depending on the related manufacturing process. Two major foam types can be distinguished, namely closed cell and open cell metal foams.

2.1 Closed cell foam

Closed cell foams are characterised by segregated pores which are not linked or connected, cf. 2.1(a). These foams typically have almost exclusively spherical cells, a very good mass to stiffness ratio, low density and excellent energy absorption properties. Therefore, closed cell foams are suited for application in numerous fields within the scope of weight optimisation and impact energy absorption, e.g. automobile and aerospace sectors [7, 86, 100].

2.2 Open cell foam

Open cell metal foams are characterised by an interconnected permeable pore structure, which exhibits a high specific surface and a moderate pressure drop. Their penetrability, allows for the use as heat exchanger, and, together with the large internal surface, for the use in chemical engineering (catalysts, battery) [7, 86, 100]. Furthermore, heat transfer properties of phase change materials may be improved by embedding open cell metal foams and increasing the penetration of heat.

2.3 Production process

Over the years, a vast number of different methods for the production of metal foams have been developed. These methods can be divided into two basic groups: smelting-metallurgy production and powder-metallurgy production.

Furthermore, the methods can be distinguished by the process of pore evolution in self-evolving and preshaping methods. In self-evolving methods, the pores are formed by gas bubbles, whereby the cell arrangement and pore size distribution is based on the laws of statistics. Thus, the final foam structure is heavily controlled by the parameters of the production process, which leads to a comparatively high scattering of the material properties. In preshaping methods (e.g. precision casting), the material properties like pore size, specific gravity or porosity, can be controlled by the mould, leading to a more homogeneous structure.

Smelting-metallurgy production: in most of these methods the molten metal is foamed by injection of gas or by addition of a blowing-agent. The latter are substances, which either perform a chemical reaction or, due to high temperatures, release dissolved gas (e.g. metallic hydrides like Titanium Hydride (TiH_2)). Corresponding production processes are for instance ALOPRAS®, METCOMB®, ALCAN/HYDRO, CYMAT, COMBAL or GASAR, cf. [100].

However, smelting production which do not use blowing agents or injection gas, are based on a modified casting process, and by utilising a mould, they belong to the group of preshaping methods. Among others, the production of open pored metal foams is commonly performed by pressure casting with placeholders, precision casting (lost-wax process) or the ERG DUOCEL®-process.

Powder-metallurgy production: this process is similar to the production of ceramics. Metal powder and the blowing-agent are mixed and compressed (extrusion pressing). The moulded piece is then brought to foaming temperature, which is close to melting temperature, where the blowing agent releases gas and a cellular structure results. These methods are not suitable for the production of open cell metal foams [100].

2.4 Characterisation of open cell foams

2.4.1 Pore density

The characterisation and classification of open-cell metal foams in terms of the pore density (ppi²) is not sufficiently accurate for the thermal and fluid mechanical studies, due the lack of information on the ratio of pore size and edge thickness. A certain number of pores per inch

---

²pores per inch

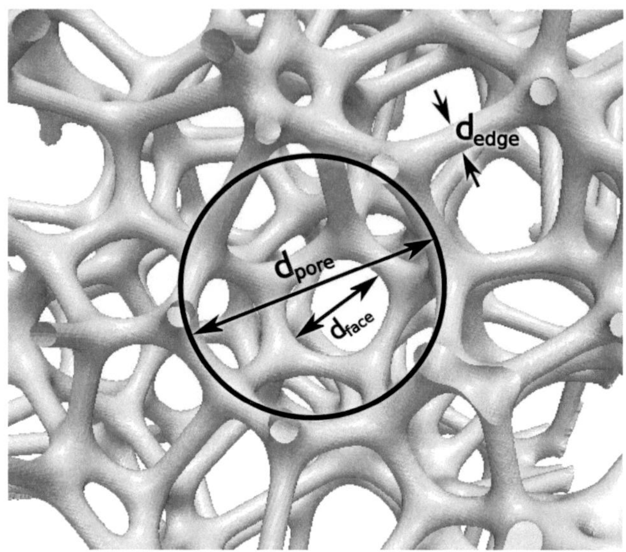

Figure 2.2: Characteristic measures of foams.

can be achieved by both, small pores with thick struts, as well as by
large pores with small struts. Despite having the same pore density, this
results in very different ratios of void (superficial) volume and volume
of the parent metal of the foam. Due to this, a characterisation of the
metal foam samples with regard to their specific gravity and the porosity
is significant for the subsequent investigations, gaining more detailed
information on the structural composition and on the permeable volume.

2.4.2 Specific gravity

The specific gravity ρ^* is defined as ratio of the density of the foam ρ_{foam}
and density of the base material ρ_{metal} [88, 89] as

$$\rho^* = \frac{\rho_{\text{foam}}}{\rho_{\text{metal}}}.\tag{2.1}$$

The more cavities or pores the foam has, the lower the specific gravity.

2.4.3 Porosity

The porosity ψ is the ratio of the cross-sectional area of the pores and the cross-sectional area of the flow channel. According to [17] the porosity is defined as

$$\psi = \frac{A_{\text{pore}}}{A_{\text{foam}}} .\qquad(2.2)$$

Based on the volume of the pores and the total volume of the foam, the porosity is written as [4, 40]

$$\psi = \frac{V_{\text{pore}}}{V_{\text{foam}}} .\qquad(2.3)$$

Furthermore, according to [100], the porosity can be deduced from the specific gravity and the density of the base material as

$$\psi = 1 - \frac{\rho^*}{\rho_{\text{metal}}} ,\qquad(2.4)$$

hence, the porosity is directly related to the relative density. Both measures characterise the structural composition of the foams and can be utilised to observe and conclude on fluid mechanical properties.

2.5 Open cell foam modelling

Making the complex geometric structures of cellular solids available to a numerical simulation requires the representation in a digitally, machine readable format. There are a couple of approaches for modelling foam structures, for example using representative elementary volumes (REV), ranging from simple cubic cell to regular dodecahedron [105] and even more complex tetrakaidecahedron [18, 42], cf. figure 2.3.

These models describe the pore structure by means of a representative unit cell, characterising the structural configuration of the foam. The unit cell is a combination of nodes which are interconnected by edges. An open-cell metal foam can thus be described as a combination of nodes and edges. Within these models, mass is distributed along struts and at the knots of the regular structures, which are then used for an analytical

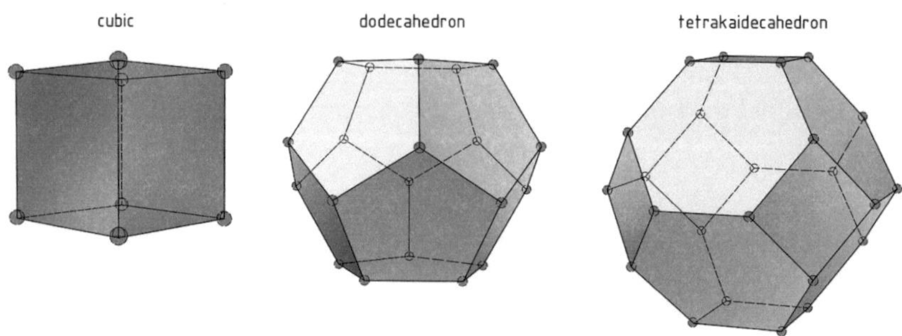

Figure 2.3: Unit cells used by different authors to synthetically build cellular solids from regular structures.

derivation of foam characteristics like porosity or structure performance data such as thermal conductivity.

In [105] the foam structure is modelled using a simple cubic elementary cell, which allows a very simple representative structure, and is used to derive a theoretical model for pressure loss and heat transfer. An even more simplified approach is used in [55], presuming one-dimensional heat transfer and modelling foam structure by means of a batch of cylinders. The cross-sectional shape of the interconnected edges is studied by [105] using the dodecahedron with twelve flat pentagonal faces. Elementary cells with prism shaped and round edges are modelled and compared in terms of porosity and pressure loss. The tetrakaidecahedron is a polyhedron with six quadratic and eight hexagonal faces [28], which allows to account for the imbalance of face and edge shape in a real foam structure. Among others it is used by [18, 42, 75, 107, 197] to model porosity, pressure loss and heat transfer in open cell metal foams.

However, detailed numerical investigations at pore scale level are either lacking in realistic structures due to simplified representation by the above mentioned unit cells, or depend on a geometry which needs to be captured by a costly tomography, cf. [12, 19, 114].

In the course of this work [178] created a new algorithm, that is capable of synthetically creating three dimensional cellular solids, foam or fabric like structures, with either open or closed cells. The workflow substantially follows three steps:

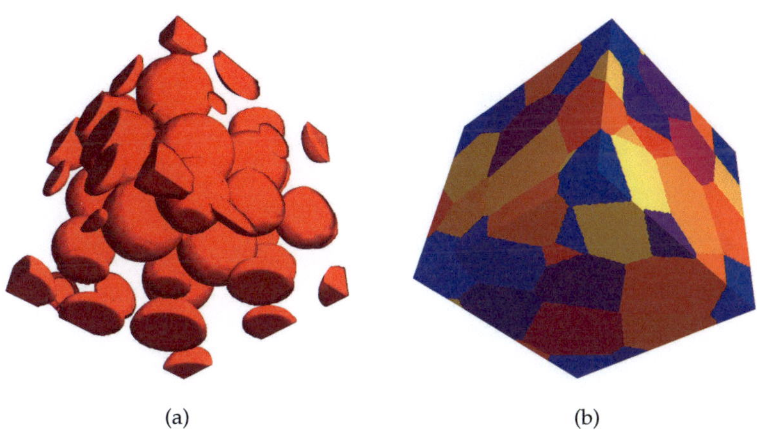

(a)   (b)

Figure 2.4: (a) Compact packing of spheres (red) with uniform pore diameter. (b) 3D Voronoi-diagram, where the different colours represent different Voronoi-regions.

1. First, the computational domain is gradually filled with spheres, in order to receive the most compact packing. Whereas the first sphere is located randomly inside the domain, the second sphere is again placed randomly at a vacant location, and then moved towards the first one. The third sphere is once more inserted at a random vacant location, and then iteratively moved towards the previous two spheres. Each following sphere is then repeatedly placed at a vacant location and then iteratively moved towards the previous three nearest-neighbour spheres. After a few iterations each sphere thus has its optimal position. Figure 2.4(a) depicts an example of a compact packing of spheres achieved by the described procedure.

2. Next, the basic topology of the structure is derived from a Voronoi-decomposition of the spatial domain, where each Voronoi-vertex corresponds to exactly one Voronoi-region. Each Voronoi-region consists of all points whose distance to the center of the Voronoi-region is not greater than their distance to any other region. Here, the Voronoi-decomposition is done, by assigning one sphere to

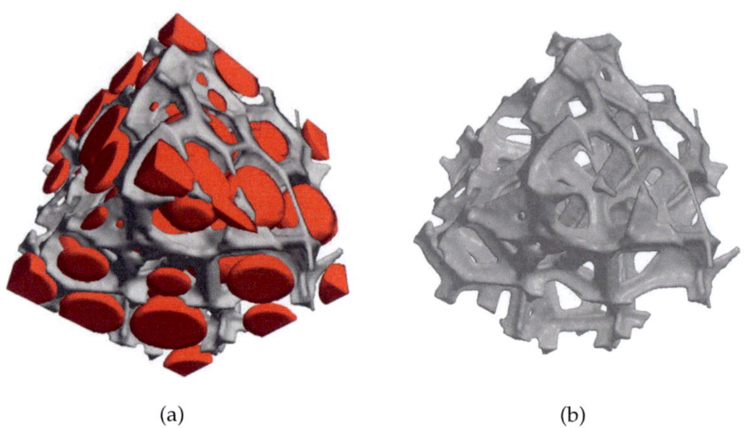

(a) (b)

Figure 2.5: (a) The struts computed from the Voronoi-diagram, (b) the final structure.

each geometrical location of the domain, for which the distance to sphere-center is minimal. The resulting Voronoi-regions give the basic topology for the cellular structure. Figure 2.4(b) shows the corresponding Voronoi-diagram for the packing of spheres given previously.

3. Finally, the cellular structure is created from the boundaries of the Voronoi-diagram. Each location, where at least three regions are connected, are used to built edges of the cellular structure. The knots of the structure, where four struts are joined, are built from the connections of four regions, cf. fig. 2.5(a).

The struts of cellular structures are often different with regards to their cross-sectional shape. For example, aluminium foams often have a triangular cross-sectional shape and thickening in the vicinity of the knots. The latter can be re-sampled by relating the thickness of the struts to the distance to their knots. The triangular shape is re-sampled, by also considering the connections of two Voronoi-regions. Here, an additional criterion if material is set or not, is given by the distance to the nearest boundary region, see fig. 2.5(b).

Chapter 3

Experiments[1]

3.1 Specimens

In the course of this work open cell metal foam is examined [202], where
the samples are aluminium foams (AlSi7Mg) with pore densities of
10 ppi, 20 ppi, and 30 ppi, and copper foam with a pore density of 10 ppi.
The probes are manufactured by precision casting and were supplied by
m.pore[2].

Figure 3.1: Exemplary open cell metal foam probes used for the experimental
investigations [67].

The cylindrical structure of the foam samples with a inner diameter
of 40 mm and a foam-length of 20 mm is embedded in the cylindrical
shroud. The shroud acts as the seal against the environment and is used
as transmission surface between the heat source and the interior metal

---

[1]Some of the results of the subsequent sections were presented at *ECCOMAS 2012 -
European Congress on Computational Methods in Applied Sciences and Engineering* and
published in [67].

[2]m.pore GmbH, Dresden, Germany

foam, which is physically connected, i.e. casted together in one single process, cf. fig. 3.1. This ensures the best possible heat transfer between the cylindrical heat sources and the foam samples.

Referring to the manufacturing process of open cell metal foam, the overall length of a cylindrical encapsulated sample is limited due to the demoulding process. With respect to the scheduled porosities for investigations, namely 10 ppi, 20 ppi and 30 ppi, the maximum sample length is limited to 20 mm. Therefore, the foam samples as well as the isolated housing are designed modularly with respect to different sample lengths planned for measurements. In doing so we presume that heat transfer effects in axial direction are negligible compared to the radial conduction and convective effects, and segmentation of foam samples still provide valuable results. Finally, specimens with a diameter of 40 mm and the incremental lengths of 20 mm, 40 mm, 60 mm, 80 mm and 100 mm are available.

3.2 Characterisation of specimens

To characterise the specimens beyond of the pore density, and to be able to better interpret the results of the fluid mechanical and thermal investigations, the samples were examined by the following described methods. The recorded measures are pore diameter d_{pore}, face diameter d_{face} and edge diameter d_{edge}, depicted schematically in fig. 2.2.

In order to assess whether the measured values are normally distributed, and if the calculated mean is representative, a normal probability plot was chosen for presentation. The so called p-value of the *Anderson-Darling* test [192] is used here as a key-figure to assess normality. In addition, the bounds of a 95 % confidence interval are shown (dotted lines) in the following diagrams.

3.2.1 Microscopy

Pore diameter d_{pore}, face diameter d_{face} and edge thickness d_{edge} are ascertained by means of a digital microscope Keyence VHX-600, by superimposing circular and linear measures using the accompanying image measurement software. Figure 3.2 shows the exemplary measurement of some circular and linear measures for a 10 ppi aluminium foam sample.

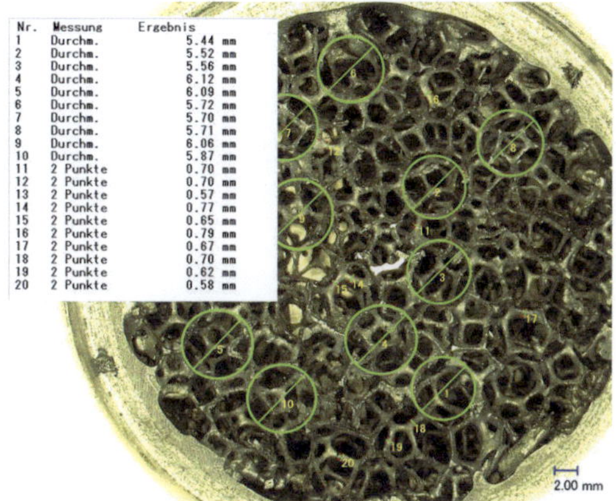

Figure 3.2: Example of superimposed circular and linear measures using the image measurement software of the Keyence VHX-600 digital microscope.

Due to the coarse structure of the 10 ppi foam samples, ten measuring points are taken on both sides of the samples at five-times magnification, making up a total of 100 measurement points. For the 20 ppi and 30 ppi samples it was possible to acquire fifteen measuring points at ten-times magnification on each side of the samples, resulting in overall 150 measurement points for each foam type.

Pore diameter

The normal probability plot of one sample for each foam type is exemplarily depicted in figs. 3.3, whereas the complete series of individual measures is given in figs. A.1 to A.4 in appendix A. It is verified by the individual normal probability plots given in appendix A for each sample of all foam types, that the assumption of normality and the averaged mean measures are valid.

On the other hand, from the cumulative normal probability plots, where all samples of one foam type are considered, we see that the criterion of the Anderson-Darling test is infringed[3]. The cumulative

---

[3]For the Anderson-Darling test, a value of $p < 0.05$ indicates a significant deviation from the prescribed probability distribution – here, normal distribution.

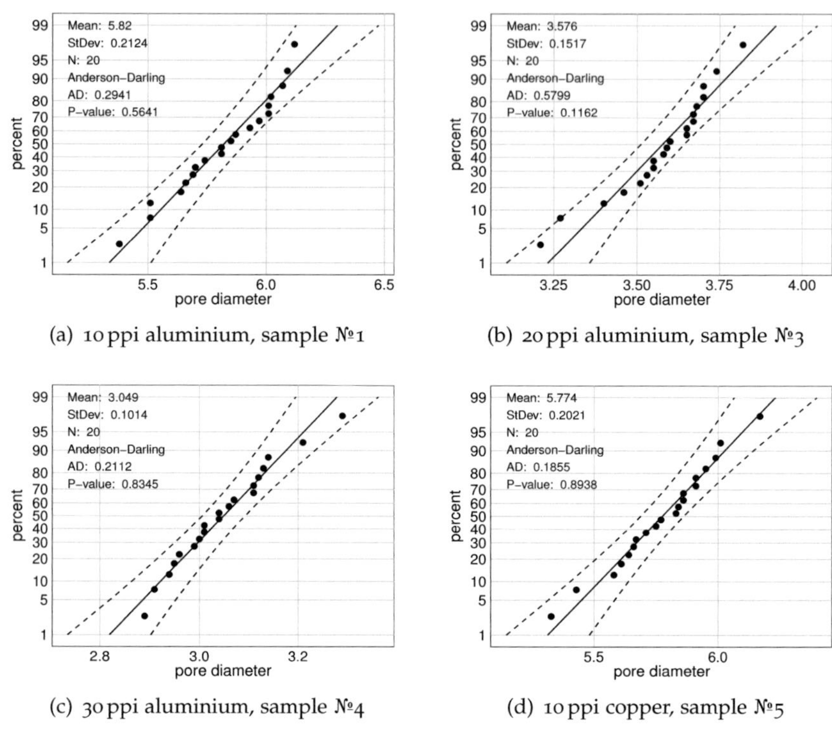

(a) 10 ppi aluminium, sample №1

(b) 20 ppi aluminium, sample №3

(c) 30 ppi aluminium, sample №4

(d) 10 ppi copper, sample №5

Figure 3.3: Exemplary normal probability plots of the measured pore diameter for (a) 10 ppi aluminium sample №1 (b) 30 ppi aluminium sample №3 (c) 20 ppi aluminium sample №4 and (d) 10 ppi copper sample №5. The dotted lines represent the 95 % confidence interval, respectively.

normal probability plots of each foam type are given in figs. A.1(f), A.2(f), A.3(f) and A.4(f), respectively.

As expected, the pore diameter decreases with increasing number of pores per inch, whereas the difference between 10 ppi aluminium and 20 ppi aluminium is much larger than between 20 ppi aluminium and 30 ppi aluminium. Furthermore, it is noteworthy, that the 10 ppi aluminium and copper foam both have a similar pore diameter of about 6 mm. The scattering of measures is highest for the 10 ppi aluminium

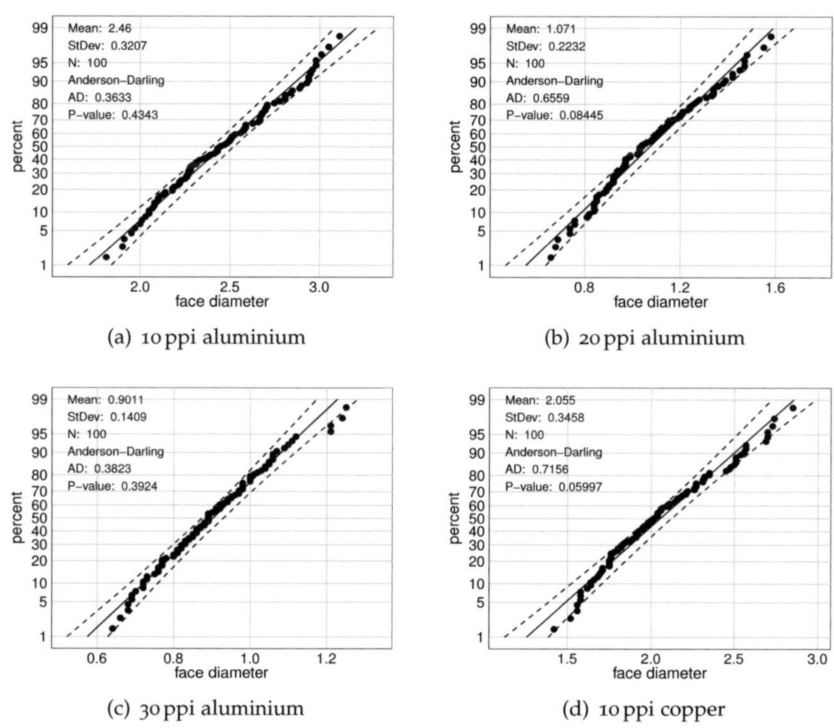

(a) 10 ppi aluminium

(b) 20 ppi aluminium

(c) 30 ppi aluminium

(d) 10 ppi copper

Figure 3.4: Cumulative (all samples) normal probability plots of the measured face diameter for (a) 10 ppi aluminium, (b) 30 ppi aluminium, (c) 20 ppi aluminium and (d) 10 ppi copper. The dotted lines represent the 95 % confidence interval, respectively.

foam. The individual mean pore diameters of each sample are summarised in table 3.1.

Face diameter

The cumulative (all samples) normal probability plots of the measured face diameters of each foam type are depicted in figs. 3.4, whereas the complete series of individual measures is given in figs. A.5 to A.8 in

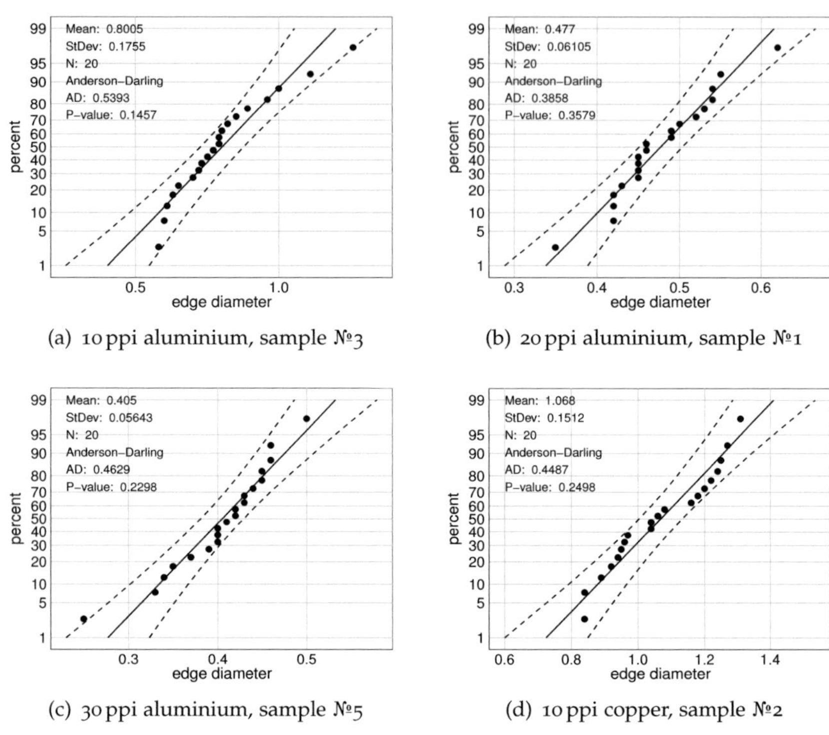

(a) 10 ppi aluminium, sample №3

(b) 20 ppi aluminium, sample №1

(c) 30 ppi aluminium, sample №5

(d) 10 ppi copper, sample №2

Figure 3.5: Exemplary normal probability plots of the measured edge thickness for (a) 10 ppi aluminium sample №3 (b) 30 ppi aluminium sample №1 (c) 20 ppi aluminium sample №5 and (d) 10 ppi copper sample №2. The dotted lines represent the 95 % confidence interval, respectively.

appendix A. Since $p > 0.05$ for the evaluation of all foam types, the face diameter does not depend on the individual samples as the pore diameter above. The individual and cumulative mean face diameters are summarised in table 3.1.

As with the pore diameter, the face diameter decreases with increasing number of pores per inch, whereas the difference between 10 ppi aluminium and 20 ppi aluminium is much larger than between 20 ppi aluminium and 30 ppi aluminium. The individual mean face diameters

Table 3.1: Summary of the individual and cumulative pore diameters d_{pore}, face diameters d_{face} and edge thickness d_{edge} for all foam samples in mm.

sample №	1	2	3	4	5	cumula-tive
10 ppi aluminium						
d_{pore}	5.82 ± 0.21	5.89 ± 0.27	6.51 ± 0.38	6.01 ± 0.29	6.32 ± 0.32	*
d_{face}	2.50 ± 0.32	2.39 ± 0.28	2.46 ± 0.35	2.52 ± 0.34	2.44 ± 0.33	2.46 ± 0.32
d_{edge}	0.60 ± 0.09	0.61 ± 0.08	0.80 ± 0.18	0.74 ± 0.12	0.78 ± 0.11	*
20 ppi aluminium						
d_{pore}	3.44 ± 0.25	3.55 ± 0.13	3.58 ± 0.15	3.57 ± 0.07	3.64 ± 0.16	*
d_{face}	1.07 ± 0.24	1.03 ± 0.19	1.05 ± 0.25	1.16 ± 0.26	1.04 ± 0.18	1.07 ± 0.22
d_{edge}	0.48 ± 0.06	0.47 ± 0.08	0.51 ± 0.01	0.48 ± 0.08	0.51 ± 0.06	0.49 ± 0.08
30 ppi aluminium						
d_{pore}	3.09 ± 0.10	3.02 ± 0.11	3.16 ± 0.15	3.05 ± 0.10	3.10 ± 0.07	*
d_{face}	0.89 ± 0.15	0.90 ± 0.14	0.93 ± 0.16	0.86 ± 0.12	0.92 ± 0.13	0.90 ± 0.14
d_{edge}	0.31 ± 0.05	0.42 ± 0.05	0.31 ± 0.05	0.43 ± 0.11	0.41 ± 0.06	0.37 ± 0.09
10 ppi copper						
d_{pore}	5.89 ± 0.46	5.69 ± 0.28	5.92 ± 0.45	5.78 ± 0.24	5.77 ± 0.20	*
d_{face}	2.06 ± 0.30	2.11 ± 0.36	1.98 ± 0.32	1.99 ± 0.33	2.14 ± 0.41	2.06 ± 0.35
d_{edge}	1.24 ± 0.11	1.07 ± 0.15	1.17 ± 0.15	1.14 ± 0.15	1.06 ± 0.29	*

*: no cumulative values available

of each sample and the cumulative face diameters are summarised in table 3.1.

Edge thickness

As with the above pore diameters, the edge thickness of the cumulative measures of the different foam types does not follow a normal distribution, whereas the normal probability plots of the individual samples given in figs. A.9 to A.12 in appendix A, probate the mean edge thickness evaluated.

While the cumulative edge thickness of the 10 ppi aluminium and copper foams infringe the criterion of the Anderson-Darling test, the cumulative values of 20 ppi and 30 ppi aluminium samples are valid. The edge thickness decreases with increasing number of pores per inch. While the pore diameter of the 10 ppi foams of aluminium and copper type are relatively similar, the edge thickness significantly differs. The individual mean edge diameters of each sample and the cumulative

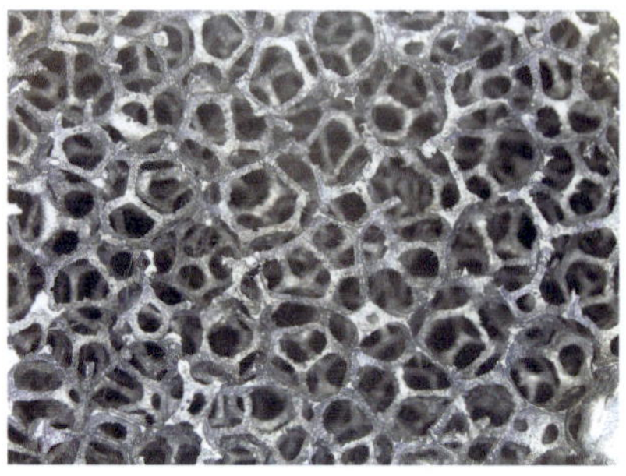

(a) 10 ppi aluminium, sample №1

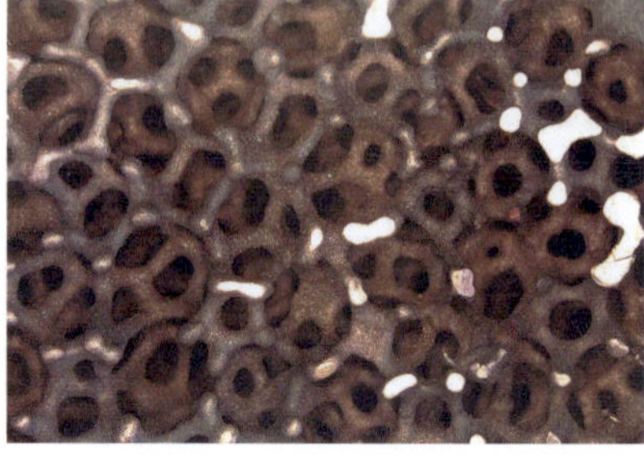

(b) 10 ppi copper, sample №4

Figure 3.6: Comparison of foam structure at ten-times magnification for 10 ppi aluminium sample №1 and copper sample №4.

(a) 10ppi aluminium, sample №1

(b) 10ppi copper, sample №2

Figure 3.7: Comparison of foam structure at twenty-times magnification for 10 ppi aluminium sample №1 and copper sample №2.

values for the 20 ppi and 30 ppi aluminium foams are summarised in table 3.1.

In direct comparison between the aluminium and copper foams of equal pore density (10 ppi) at ten-times magnification distinct differences in the pore diameter, face diameter and edge thicknesses can be seen, cf. fig. 3.6. While the copper foam has smaller pore and face diameter, the edge thickness is larger, thus achieving the same pore density, such as aluminium foam. The difference in edge-shape of aluminium and copper

Table 3.2: Summary of the individual and cumulative specific gravity ρ^* and porosities ψ.

sample №	1	2	3	4	5	cumulative
10 ppi aluminium						
ρ^*	0.116	0.072	0.123	0.162	0.121	0.119
ψ	88.441%	92.777%	87.703%	83.846%	87.888%	88.131%
20 ppi aluminium						
ρ^*	0.170	0.166	0.210	0.171	0.169	0.177
ψ	83.032%	83.355%	78.996%	82.926%	83.111%	82.284%
30 ppi aluminium						
ρ^*	0.126	0.177	0.090	0.189	0.127	0.141
ψ	87.439%	82.334%	90.999%	81.143%	87.341%	85.851%
10 ppi copper						
ρ^*	0.236	0.298	0.289	0.239	0.272	0.267
ψ	76.402%	70.186%	71.067%	76.074%	72.834%	73.312%

foams is depicted in figs. 3.7 at twenty-times magnification. While the aluminium foam shows triangular shaped edges, the edges of the copper foam are rounded. Due to the figs. 3.6 and 3.7 it can be surmised that the porosity of the copper foam is smaller than that of the aluminium foam.

3.2.2 Gravimetry

Since the foam samples are embedded into the shroud, a direct measurement of the mass of the foam is not possible. Instead, the finally machined geometry of the shroud is measured in detail, and the mass was calculated using the density of the base material specified by the manufacturer and subtracted from the overall mass of the samples.

The overall weight of the individual samples is measured using a Sartorius digital precision scale type LP6200S. All measurements are repeated multiple times, and all results are controlled using a Mettler analogue precision scale type P-1200.

Doing so, the foam mass is calculated from $m_{\text{foam}} = m_{\text{sample}} - m_{\text{shroud}}$, whereas the specific gravity and the porosity are evaluated using eqns. (2.1) and (2.4), respectively. The specific gravity and the porosities of the individual samples are summarised in tab. 3.2.

Even though the specimens are delivered by the same manufacturer and originate from the same production lot, there are significant differ-

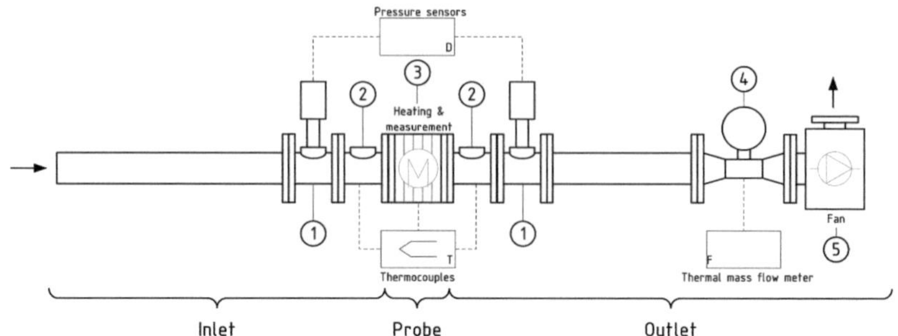

Figure 3.8: Schematic diagram of the test rig: inlet, probe and outlet with pressure traversing units ①, temperature rakes ②, probe section ③, thermal massflow meter ④ and air supply ⑤.

ences in porosity. The 30ppi aluminium samples show the maximum deviation in porosity, varying in the range between 81.144% and 90.999%.

This again illustrates the shortcoming of the term pore-density (ppi – pores per inch). Foams of the same pore density significantly differ regarding their pore scale measures and porosities, whereas foams of different pore densities show similar porosity.

3.3 Test rig

The experimental test rig for fluid flow and heat transfer measurements of air flow through cellular solids is designed, manufactured and assembled at the Karlsruhe University of Applied Sciences [64, 67]. Emphasis is laid on practical relevance, and a circular cross-section is chosen in the style of most engineering fluid flow applications. Furthermore, this avoids impact of secondary flow phenomena.

The cross sectional area with a diameter of 40 mm is designed with respect to a feasible flow supply and to enable laminar up to turbulent flow regime with respect to the available metal foam probes of different porosities. According to the specifications of the manufacturer, probes with 10 ppi, 20 ppi and 30 ppi are customly produced.

The test rig can be divided into three sections, namely the inflow-, test- and outflow-section, where the temperatures, static and total pressures are measured at two locations at the inlet and outlet section, right in front and behind of the test section, cf. fig. 3.8.

In the inflow section of the experimental rig, ambient air flows through a stainless steel pipe with a length of 2 m, such that the flow is unimpeded and fully developed. In order to minimise the axial offset of the pipings and fittings, either metal bushes are used for alignment, sealed and mounted using TKA-Axilock couplings, or grooved flanges. In order to measure the pressure and velocity profiles in front of and behind the foam samples, a custom made adjustable Pitot tube is designed and manufactured, cf. figs. 3.9. It enables to measure the total pressure along the transversal axis of the cross section at any position up to a distance of 2 mm from the wall. A stainless steel capillary tube with an outer diameter of $D = 1$ mm and a diameter ratio of $d/D \approx 0.6$ is used for the Pitot tube, cf. fig. 3.9(b). According to [159], for the diameter ratio of about 0.6, a deviation of $\leq \pm 10°$ causes no significant error. The linear actuator allows positional accuracy and repeatability by 0.01 mm. Using the static pressure measured by four wall pressure taps connected by a loop, we are able to obtain the dynamic pressure and velocity respectively. The Pitot tube is driven by a linear actuator which is controlled by the computer-operated test rig control system. Throughout all pressure taps, high accuracy capacitive and piezoresistive differential pressure transmitters of Endress+Hauser with ceramic or silicon sensors are used.

Figures 3.10 shows the custom made temperature rake, which is used to ascertain the temperature profiles in front of and behind the foam samples. The rake is equipped with 9 miniature high accuracy type T thermocouples with a diameter of 0.5 mm.

The test section is entirely made of PTFE and designed in a modular way, to enable testing of probes with lengths of 20 mm, 40 mm, 60 mm, 80 mm and 100 mm. It consists of a front and rear element and five identical intermateable slabs. The front and rear plates are designed and milled to be aligned with the grooved flanges of the measurement fittings. An aligned and tight connection between the intermateable slabs are guaranteed by a centering collar and Viton

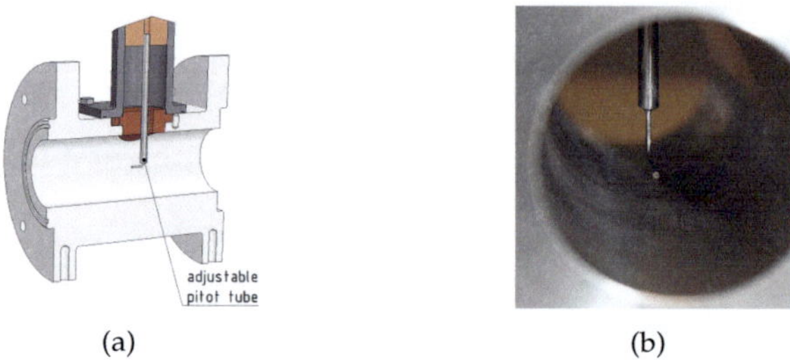

(a)   (b)

Figure 3.9: Detailed view on (a) the three dimensional design models of the custom made pressure probe section of the test rig (linear actuator and sensor are omitted here) and (b) image of the Pitot tube inside the 'real' assembly.

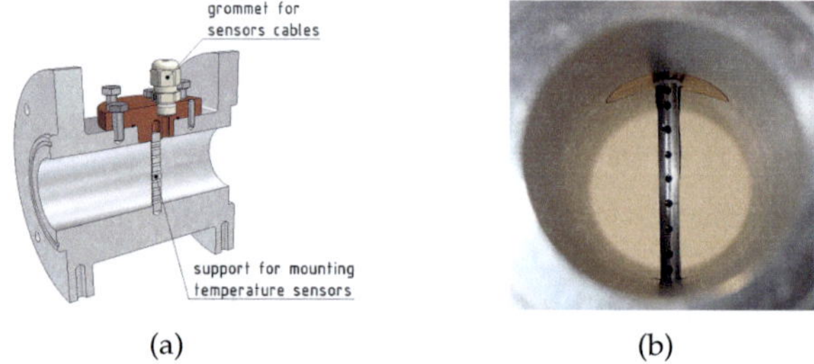

(a)   (b)

Figure 3.10: Detailed view on (a) the three dimensional design model of the custom made temperature probe section of the test rig and (b) the temperature rake inside the probe section.

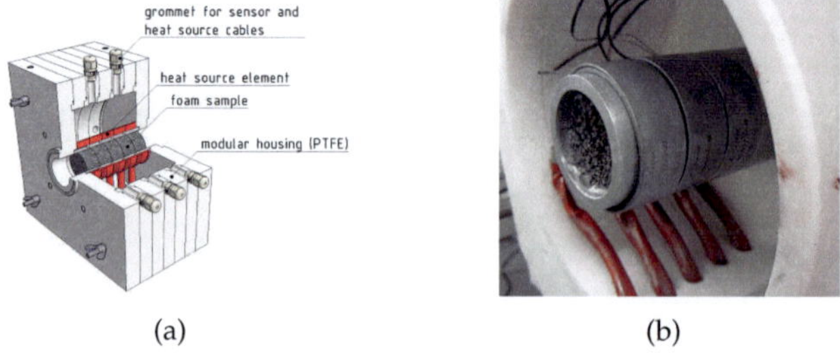

(a)   (b)

Figure 3.11: Three dimensional design model of the test section of the test rig, with foam samples, heat sources and modular housing elements; the temperature sensors are omitted here.

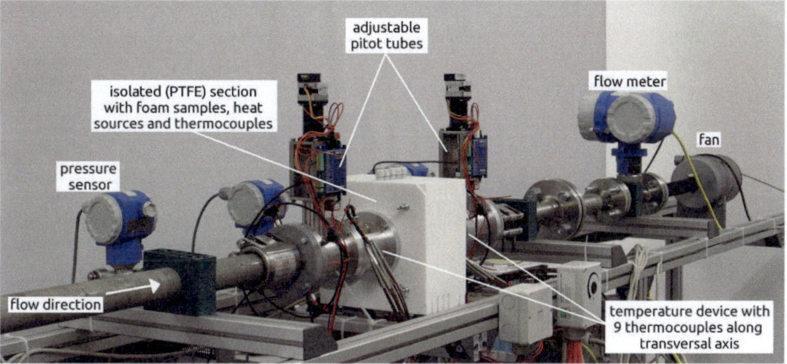

Figure 3.12: Final installation of the test rig.

o-rings. The overall test section is assembled using four threaded rods and wing nuts, cf. figs. 3.11.

For the heat transfer measurements, the foam samples are equipped individually with cylindrical heat source elements from Freek with an output of 470 W, i.e. roughly a total output of 2500 W when testing five samples. Each heating element is controlled individually by the test-rig control software, such that the wall temperature is kept constant.

The outflow section is equipped with a Endress+Hauser thermal mass flow meter Proline t-mass 65, which is customly calibrated for a limited range of 0 to 65 kg/h to obtain a measurement uncertainty of $\pm 1\,\%$. At the outlet of the open circuit an ebm-papst high-pressure blower type G3G125-AA20-10 is installed for air supply. The blower provides a minimum massflow of 5 kg/h, a maximum massflow of 65 kg/h and a maximum pressure difference of 18'000 Pa at zero flow rate.

3.4 Control unit & data acquisition

The complete governance and control system is realised with a customised National Instruments LabVIEW software control system, allowing the simultaneous recording of all signals as well as the automatic steering

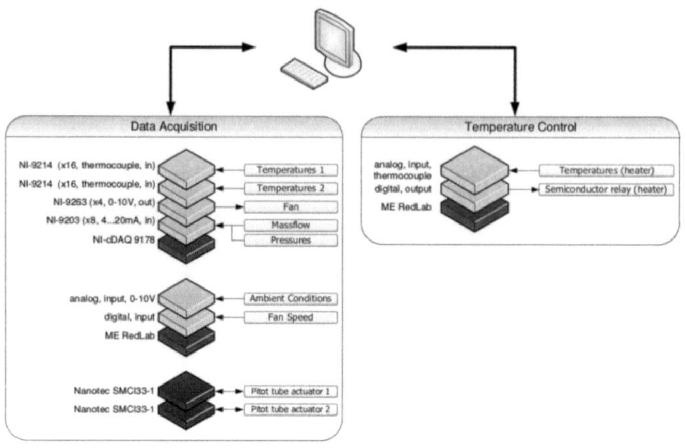

Figure 3.13: Schematic of the data acquisition, control system and regulation.

of the adjustable Pitot tube and the automatic heating control. The latter allows a constant temperature regulation at $\pm 1^\circ$C within a temperature range of about 30°C up to 120°C at full discharge.

The connection to the respective actuators and sensors is realised via a NI-cDAQ-9178 USB data acquisition board and several modules for the processing of the different input and output signals, cf. fig. 3.13.

The pressure sensors and the thermal massflow meter are connected via an analogue input module NI-9203 while there are two dedicated 16-channel input modules NI-9214 for the thermocouples, and the fan is controlled by the NI-9263 output module. The linear actuators of the Pitot probes are directly connected via USB, and also controlled within the LabVIEW program. Ambient conditions as well as the temperature control of the heat sources are connected using RedLab USB boxes from Meilhaus Electronics.

The graphical user interface of the test-rig control software allows the to user access the fan speed, the positioning of the Pitot tubes, the constant temperature regulation and the automatic acquisition of all data, cf. fig. 3.14. Due to the huge amount of data (about 60'000 values for a measurement time of 1 min at a sampling rate of 1000 Hz) the software DIAdem of National Instruments is used for post-processing.

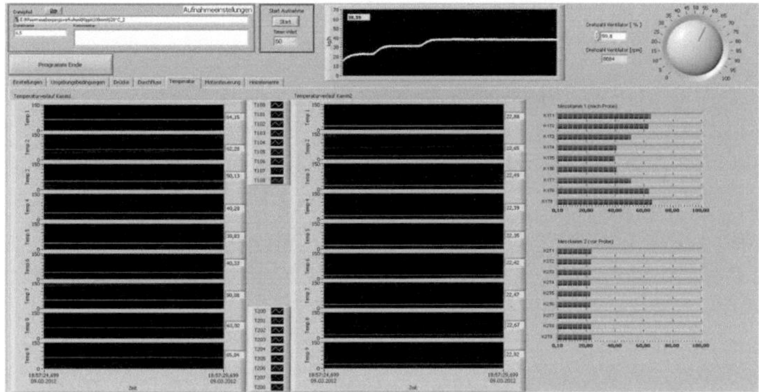

Figure 3.14: Exemplary screenshot of the test-rig control and data acquisition software program, realised with National Instruments LabVIEW software.

Table 3.3: Absolute total uncertainties for the different signals.

value	unit	total absolute uncertainty
ambient pressure	Pa	± 395
ambient temperature	°C	± 1.0
rel. humidity	°C	± 0.8
static/total pressure	Pa	± 2.5
dynamic pressure	Pa	± 6.1
massflow	kg/h	± 0.1
temperatures	°C	± 0.6

3.5 Measurement uncertainties

The measurement chain is affected by several sources of uncertainty. In addition to the uncertainties of the sensors, there are sources of uncertainties in signal conditioning (offset errors and gain errors), as well as the quantisation error of the A/D conversion. With respect to the rules of error propagation, the absolute total uncertainty for the different sensors and the relevant measurement chain are determined and given in tab. 3.3. Further details and tests on repeatability, accuracy and plausibility are reported in [202].

3.6 Fluid mechanic measurements

Knowledge about the head loss of flow in open cell metal foam is vital for its application as heat exchanger. A first attempt for the description of the flow in porous media was done in the early 18th century by Darcy. Assuming homogeneity and isotropy of the porous layer, he derived by experiments the proportional relationship between the pressure gradient ∇p, the discharge rate q and the dynamic viscosity μ of the fluid

$$-\nabla p = \frac{\mu}{K_1} q \, , \tag{3.1}$$

where K_1 is the empirical permeability of the porous medium [113]. Relating the gradient of a potential to a flux via a mobility coefficient is similar to Fick's law, Fourier's law or Ohm's law, and can also be derived as a special solution of the Navier-Stokes equation. The relation can be used for flow regimes with Reynolds numbers ≤ 3. Increasing velocity or decreasing viscosity leads to a higher Reynolds number, where inertial forces become crucial compared to frictional forces. This is taken into account in the Forchheimer-Darcy equation, by adding a turbulent term to the viscous term on the right hand side of the Darcy equation

$$-\nabla p = \frac{\mu}{K_1} q + \frac{\rho}{K_2} q^2 \, . \tag{3.2}$$

For the application of Darcy's law and the Forchheimer-Darcy equation for packed beds, the Darcy and non-Darcy permeability coefficients K_1 and K_2 are for instance given as correlations of particle diameter, porosity and Reynolds number, and known as Kozeny–Carman equation and Ergun-equation.

There are numerous publications concerned with pressure loss data and correlations for open cell metal foam as well as ceramic foams, and this issue is still subject of ongoing research activities, cf. [2, 51, 53, 59, 82, 133, 134]. The most comprehensive overview of the state of the art empirical and theoretical methods for pressure drop modelling is given by [59]. In accordance with the well known Ergun-equation for packed spheres, many authors make use of a second order polynomial for successfully modelling pressure loss correlations.

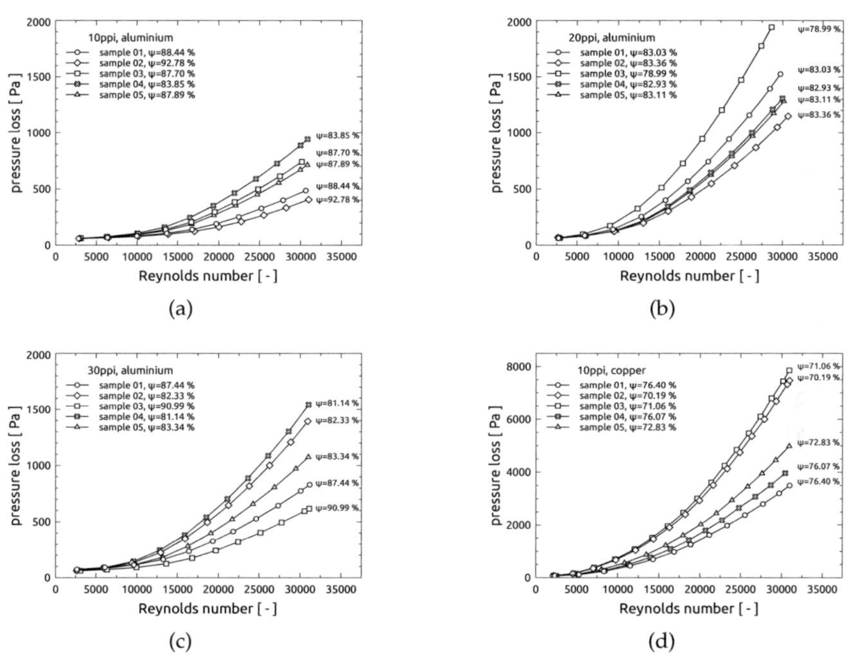

Figure 3.15: Pressure loss for all the individual foam samples of (a) 10 ppi aluminium, (b) 20 ppi aluminium, (c) 30 ppi aluminium and (d) 10 ppi copper, respectively. Reynolds number is based on inlet diameter.

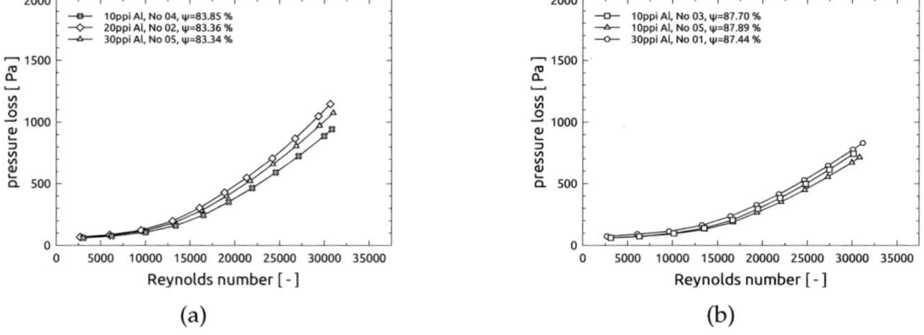

Figure 3.16: Pressure loss of foams of similar porosity (a) $\psi \approx 83\%$ and (b) $\psi \approx 87\%$. Reynolds number is based on inlet diameter.

3.6.1 Pressure loss of all individual samples

Testing all individual samples, the measurements range from minimum fan speed to the upper limit of the thermal massflow meter or 100 % fan speed, with increments of 5 %. Thus, for each sample we conduct 10 to 16 individual measurements, depending on the specific pressure loss.

Each measurement point represents the temporal average of a measurement duration of 10 s, respectively. All physical properties are determined according to the inlet conditions of the working fluid and [110]. The Reynolds number reads

$$\mathrm{Re} = \frac{u_s\, d_i}{\nu_f}\,, \tag{3.3}$$

where u_s is the superficial velocity at the inlet of the specimen, d_i is the diameter of the inlet geometry and ν_f is the kinematic viscosity of the fluid with respect to the inflow conditions.

Figures 3.15 show the pressure losses of the individual foam samples for all foam types and pore densities, respectively. The individual samples of equal pore density show significant differences in pressure loss, which corresponds to the variations in characteristic pore scale measures already established in sec. 3.2. However, the deviations are consistent with the porosities, i.e. higher porosity causes less pressure loss. The sole exceptions are 10 ppi copper samples №2 and №3 as well as 20 ppi aluminium samples №1 and №4, cf. figs. 3.15(b) and 3.15(d), respectively.

Reasons for the discrepancies are the measuring inaccuracies in gravimetric determination of porosity and material accumulation in the vicinity of the shroud, cf. figs. 3.17. As already mentioned in sec. 3.2.2 these are production-related issues, which cannot be excluded in the course of manufacturing, neither be controlled nor suppressed.

Foams of different pore densities but similar porosities are compared in figs. 3.16. Different foams of similar porosity compare well to each other, and it becomes obvious, that the pore density is neither significant nor meaningful to characterise similarity or comparability.

3.6.2 Pressure loss per unit length

In a second setup, the samples of the same pore density are combined modularly, testing the gradations of 20 mm, 40 mm, 60 mm, 80 mm and

(a) 10 ppi copper, sample №2

(b) 10 ppi copper, sample №3

(c) 20 ppi aluminium, sample №1

(d) 20 ppi aluminium, sample №4

Figure 3.17: Material accumulation in the vincinity of the shroud as well as measurement inaccuracies are the reason for uncertainties in determination of porosity: (a) and (b) 10 ppi copper foam samples №2 and №3, (c) and (d) 20 ppi aluminium foam samples №1 and №4, respectively.

100 mm, respectively. Starting at minimum fan power, the measurements are limited by either the maximum fan power or the upper limit of the thermal massflow meter, depending on porosity and combined sample length. As established above, the porosity varies significantly for samples of the same pore density, which already became evident in the pressure loss characteristics of the individual samples. We therefore expect a certain scattering for the pressure loss per unit length.

The results of the measurements are given in figs. 3.18, together with a quadratic fit of type $y = a\,x^2 + b\,x$ for each pore density, respectively. The 10 ppi copper foam shows the largest scattering, and remarkably

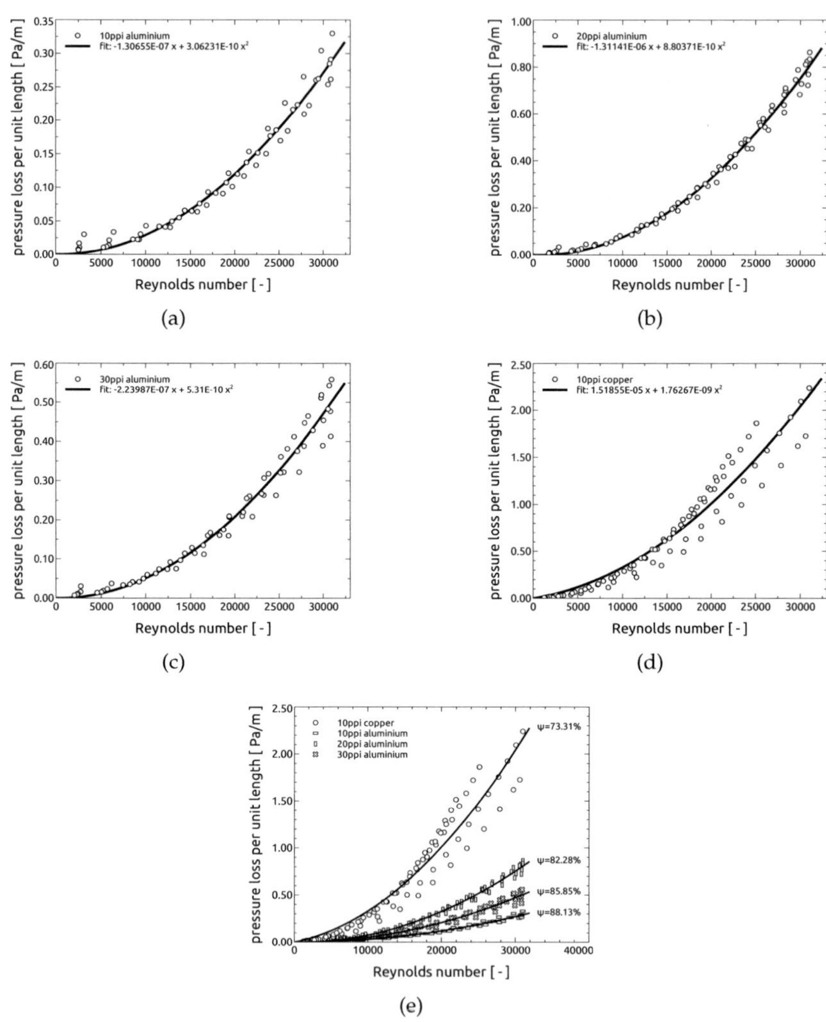

Figure 3.18: Pressure loss per unit length for (a) 10 ppi aluminium, (b) 20 ppi aluminium, (c) 30 ppi aluminium, (d) 10 ppi copper and (e) comparison of all foam types. The black solid line represents a quadratic fit of type $y = a\,x^2 + b\,x$. Samples of the same pore density are combined for overall lengths from 20 mm to 100 mm.

good results are obtained for the 20 ppi aluminium foams. From this it is evident, that the pore density is unsuitable for the characterisation of foams, even though the individual samples of one pore density are manufactured by the same production batch. Figure 3.18(e) shows the comparison of the pressure loss per unit length for all pore densities. Despite of the scattering, the ordering of the pressure loss characteristics corresponds to the average porosities of the samples.

In order to correlate the experimental data of all samples, the pressure loss is described by the dimensionless Hagen number and the flow velocity by the Reynolds number as

$$\mathrm{Hg} = \frac{\Delta p}{\Delta L} \frac{d_h^3}{\rho_f v_f^2} \qquad \text{and} \qquad \mathrm{Re} = \frac{u\, d_h}{\psi\, v_f} \tag{3.4}$$

Following the *Carman-Kozeny* theory, and according to the hydraulic radius model [113], the hydraulic diameter is expressed as

$$d_h = \frac{4 \times \text{void volume}}{\text{surface area}} = \frac{4\,\psi}{S_v(1-\psi)}, \tag{3.5}$$

where S_v is the specific surface area. The latter can be derived from the pore diameter d_{pore}, the face diameter d_{face} and the porosity ψ by the correlation proposed by [22] based on a tetrakaidecahedron model

$$S_v = \frac{C_v(1-\psi)^n}{d_{\mathrm{edge}} + d_{\mathrm{face}}}, \tag{3.6}$$

with the constants $C_v = 0.482$ and $n = 0.5$.

The non-dimensional pressure loss per unit length and the non-dimensional velocity of all foam types and lengths is given in fig. 3.19, in terms of Hagen and Reynolds number, eqns. (3.4), respectively. The Reynolds number varies in the range of about $350 < \mathrm{Re} < 35\,000$, where the Hagen numbers of all samples can be represented by a correlation of type $\mathrm{Hg} = A \cdot \mathrm{Re} + B \cdot \mathrm{Re}^2$ with respect to Erguns equation, and in accordance to [47]. The dashed line represents the correlation of [47], obtained from measurements on different ceramic foams, whereas the solid line represents the correlation found for the measurements done in the course of this work. Here, the coefficient of the linear term is

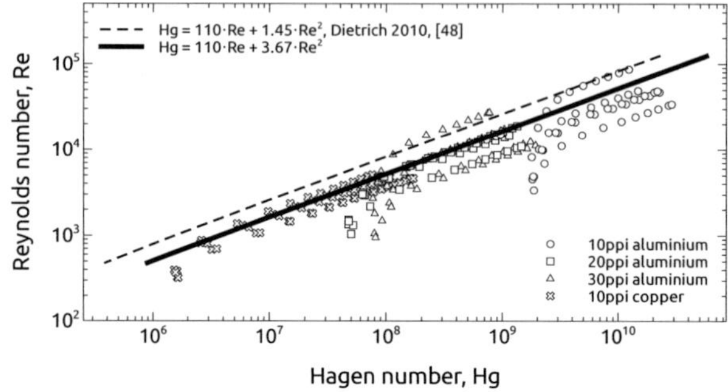

Figure 3.19: Correlation of the non-dimensional pressure loss per unit length and non-dimensional velocity, in terms of Hagen and Reynolds number for all foam types and lengths.

retained constant, since it mainly influences the slope of the correlation curve towards low Reynolds numbers $Re \leqslant 100$, which are not subject of the present measurements. In the work of [47], the maximum Reynolds number is at about $Re \approx 3'000$, whereas in the present study the foam samples are investigated for Reynolds numbers up to $Re \approx 35'000$. With respect to Erguns equation, the coefficient of the quadratic term models the contribution due to turbulent effects, which are more crucial for the mentioned flow regime. The quadratic coefficient B is found by a least square fit, with respect to the double logarithmic reference frame. The correlation is in good agreement with the measured data. However, for each foam type, the records at the lower end of the Reynolds number show higher deviations compared to the correlation.

In Accordance with measurements in [17], these are the transition points of the flow regime, i.e. the pressure drop across the metal foam leaves the linear Darcy regime and enters the form dominated pressure drop regime.

In [181] similar deviations are reported for the heat transfer at low Peclet numbers in packed beds. It is assumed, that the low Reynolds number deviations are subject to similar effects, where flaws in the cellular structure, e.g. closed cells or the material accumulation at the shroud of the samples (cf. figs. 3.17) cause imbalances in the flow field which result in modified pressure loss characteristics.

3.7 Thermal measurements

The use of high conductivity cellular solids has gained increasing importance in heat transfer applications over the recent years. Open cell metal foams provide the advantages of metallic properties like good heat conductivity with a large surface to volume ratio and an intense mixing of the flow. Numerous publications, on experimental and theoretical work carried out in the recent past [20, 52, 55, 85, 125, 126, 132, 199, 218], are evidence for the importance of developing a better foundation and even a better understanding with respect of the thermal design.

3.7.1 Mechanisms of heat transfer

Most commonly one distinguishes between heat transfer by conduction, by convection and by radiation, whereby the heat flux q is the rate of thermal energy flow per unit surface area which is subject to the heat transfer.

The macroscopic formulation of heat conduction is known as Fourier's law, where the heat flux is proportional to the temperature gradient

$$q = -k\nabla T, \tag{3.7}$$

where the proportionality factor k is the thermal conductivity, and the minus sign indicates, that the direction of heat transfer coincides with the negative temperature gradient.

Heat transfer in terms of the transport by a fluid flow is subject to convective heat transfer. One can distinguish between natural or free convection, if the fluid flow is driven by the buoyancy forces induced by the temperature gradients, or forced convection, if the fluid flow is driven by a pump. However, such energy transfer heavily depends on the nature of the flow [8, 122]. With regards to Fourier's law, within the linear Ansatz of heat transfer the convective heat transfer is characterised by a heat transfer coefficient

$$h = \frac{|q|}{\Delta T}, \tag{3.8}$$

given as the proportionality of the magnitude of the heat flux and the driving force of the heat flux, i.e. the characteristic temperature difference.

Thermal radiation takes into account the emission and absorption of electromagnetic radiation due to an object's temperature. For the temperature ranges covered in the experiments, thermal radiation cannot be generally excluded. Therefore, heat transfer considered in the following represents the sum of convection and radiation effects.

3.7.2 Characterisation of heat transfer

In general, heat transfer can be characterised by the non-dimensional representation of the heat transfer coefficient h, namely the Nusselt number

$$\mathrm{Nu} = \frac{h\, l_{\mathrm{ref}}}{k_f}, \tag{3.9}$$

where l_{ref} is a characteristic length scale and k_f is the thermal conductivity of the fluid.

The local heat flux q_w at the surface of the foam structure depends on the local flow and thermal boundary layer, and can hardly be measured. Therefore, in the following an integral heat transfer coefficient is used

$$\bar{h} = \frac{\dot{Q}}{A_{\mathrm{ref}}\, \Delta\vartheta}, \tag{3.10}$$

where $\dot{Q}$ is the rate of heat flow, A_{ref} is the characteristic heat transfer surface and $\Delta\vartheta$ is the temperature difference. Analogous to a channel-flow, the integral rate of heat flow is defined as

$$\dot{Q} = \dot{m}\, c_P \left(\vartheta_{fo} - \vartheta_{fi}\right) \tag{3.11}$$

with $\dot{m}$ the massflow of the fluid, c_P the specific heat capacity at constant pressure of the fluid, and $\vartheta_{fo} - \vartheta_{fi}$ the temperature difference of the fluid between inlet and outlet. With respect to the constant wall temperature, the temperature difference $\Delta\vartheta$ in eq. (3.10) is replaced by the logarithmic temperature difference [170]

$$\Delta\vartheta_{log} = \frac{\left(\vartheta_{fo} - \vartheta_{fi}\right)}{\ln\left(\dfrac{(\vartheta_W - \vartheta_f)_i}{(\vartheta_W - \vartheta_f)_o}\right)}. \tag{3.12}$$

Substituting eq. (3.10) with definitions in eqns. (3.11) and (3.12) in the definition of the Nusselt number (3.9) reads [132]

$$\mathrm{Nu} = \frac{\dot{m}\, c_P\, l_{\mathrm{ref}}}{A_{\mathrm{ref}}\, k_f} \ln \left(\frac{(\vartheta_W - \vartheta_f)_i}{(\vartheta_W - \vartheta_f)_o} \right). \tag{3.13}$$

Depending on the level of approximation which applies for the system under investigation, different measures apply for the characteristic length [113]. Looking at the metal foam as a whole [18, 40], the hydraulic diameter d_{hydr} is used in the definition of the Reynolds number, whereas the overall foam length l_{ref} is used in the definition of the Nusselt number. In accordance with the approach used for porous beds other authors replace the particle diameter with the pore diameter d_{pore}, or the permeability K, cf. [54, 55, 75].

Since the temperature and pressure sensors are applied right in front of and behind the specimens, the characteristic values represent integral measures at the macroscopic level. Therefore, and in accordance with [18, 40], the characteristic lengths d_{hydr} and l_{ref}, are used for the Reynolds and Nusselt number, respectively. The cylindrical duct wall is used as characteristic heat transfer surface A_{ref}.

The utilisation of individual samples, used to increase the overall sample length by discrete increments, is done under the assumption of negligible axial heat conduction. The latter, in terms of a non-dimensional transport equation of the temperature reads

$$\frac{1}{Pe^2} \frac{\partial^2 T}{\partial \zeta^2}, \tag{3.14}$$

where ζ is the non-dimensional axial coordinate, and Pe is the Peclet number [47]. Obviously, for $Pe \gg 1$, the impact of axial heat conduction is negligible. For the properties and conditions applied in this work, with $Pe > 1000$, it is reasonable to conclude that heat conduction in axial direction can be neglected.

3.7.3 Execution of tests

Heat transfer measurements are carried out at varying massflows and with constant wall temperatures of 60°C, 90°C and 120°C. For all of

Table 3.4: Coefficient C and exponent m for the individual Nusselt number correlations of type $\mathrm{Nu} = C \cdot \mathrm{Re}^{m} \cdot \mathrm{Pr}^{1/3}$, according to [179, 218].

foam type	Al 10 ppi	Al 20 ppi	Al 30 ppi	Cu 10 ppi
length, l [mm]		exponent, m		
100	0.651	0.706	0.699	1.071
80	0.631	0.651	0.687	1.045
60	0.636	0.600	0.620	0.886
40	0.586	0.601	0.615	0.711
20	0.650	0.515	0.581	0.507
length, l [mm]		coefficient, C		
100	4.265	3.027	2.985	0.152
80	4.130	4.189	2.795	0.165
60	3.476	5.648	4.340	0.614
40	3.970	4.158	3.842	2.445
20	1.325	4.491	2.700	7.509

the four pore densities, the measurement series are conducted using all samples, thus with an overall length of 100 mm, applying different mass flows at three different wall temperatures, 60 °C, 90 °C and 120 °C respectively. Then, the overall length is reduced by one sample (20 mm) continually, each time measuring different mass flows at three wall temperatures.

About 80 measurement series are carried out, where each series includes 10 operating points in the range of 10 % to 100 % massflow. Each time, the parameters of the constant wall temperature control unit are adjusted, which allows to keep the maximum deviation within $\pm$ 0.3 °C of the index value. After the control system settles and all measurement signals have stabilised, the recording of data is done for 60 s, corresponding to 60'000 data records of temperatures, pressures and massflow.

The average temperature at the inlet and outlet of the samples is derived by means of a massflow average. However, comparison of area and massflow averaging shows a maximum difference in Nusselt number of approximately 4 %. The measurement of the massflow and velocity distribution is explained in more detail in section 3.8.

$$\langle \vartheta \rangle_{\mathrm{a}} = \frac{1}{\sum_{i=1}^{n} A_i} \sum_{i=1}^{n} A_i \vartheta_i \tag{3.15}$$

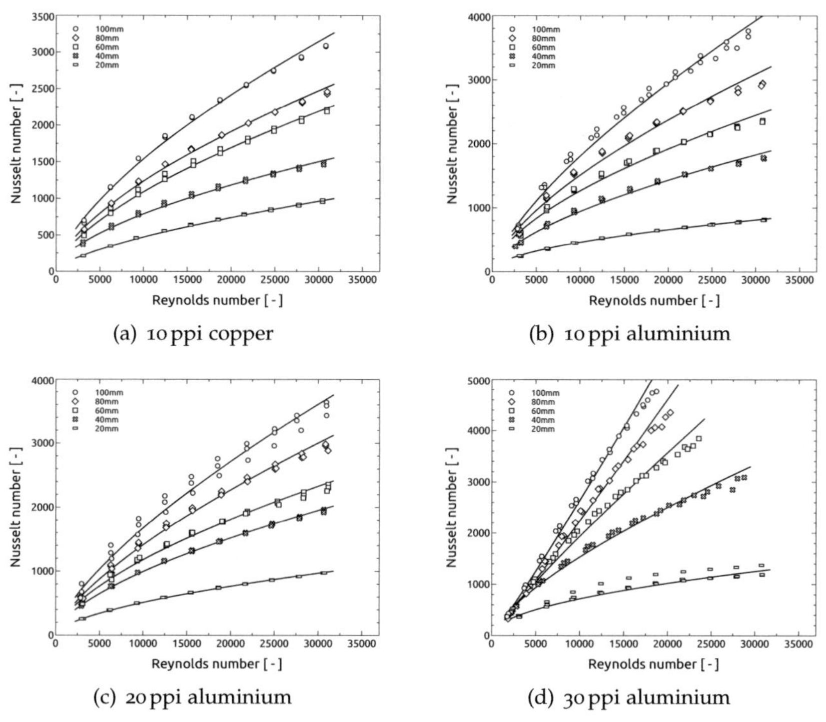

(a) 10 ppi copper

(b) 10 ppi aluminium

(c) 20 ppi aluminium

(d) 30 ppi aluminium

Figure 3.20: Nusselt number distributions of all foam types and lengths, each measured at three temperatures, 60 °C, 90 °C and 120 °C, respectively. The solid lines represent correlation curves of type $\mathrm{Nu} = C \cdot \mathrm{Re}^m \cdot \mathrm{Pr}^{1/3}$, for which the coefficients C and exponents m are individually fitted for each length; values are tabulated in tab. 3.4.

$$\langle \vartheta \rangle_{\mathrm{m}} = \frac{1}{\sum_{i=1}^{n} \dot{m}_i} \sum_{i=1}^{n} \dot{m}_i \vartheta_i \tag{3.16}$$

Regarding the material properties, the density of the air ρ_f is calculated using the equation of state with respect to the inlet conditions, whereas the specific heat capacity at constant pressure c_P, the thermal conductivity k_f and the viscosity ν_f are interpolated from tabulated data, cf. [110]. The

Reynolds number is varied from about 600 to about 30'000, considering both, laminar and turbulent flow regimes.

3.7.4 Experimental heat transfer coefficients

Figure 3.20 depicts the measured heat transfer coefficients shown as Nusselt number distributions for all foam types, for the lengths 20 mm, 40 mm, 60 mm, 80 mm and 100 mm and for the three temperature levels 60 °C, 90 °C and 120 °C. The fluid properties exhibits minor variations for the applied temperature ranges, wherefore the measurements at the three temperature levels almost coincide.

As expected, the Nusselt number increases with higher Reynolds numbers and larger sample lengths for all foam types. The Prandtl number is found to be virtually constant for the temperature range of the experiments, and is evaluated at approximately $\mathrm{Pr} \approx 0.72$.

Following [179, 218] a correlation of type $\mathrm{Nu} = C \cdot \mathrm{Re}^m \cdot \mathrm{Pr}^{1/3}$ is chosen, where the coefficient C and the exponent m are adapted for each foam type and porosity. The correlations are given by the solid lines in figs. 3.20. The coefficient C and the exponent m are tabulated in table 3.4. There is no uniform trend for the coefficient and exponent for different sample lengths, which can most probably be attributed to the deviations in porosity and specific gravity for the samples of the same pore density, cf. sec. 3.2. However, the individual fitted correlations are in good agreement with the experimental data.

With respect to [26, 179], employing the same correlation but utilising a constant exponent m for each pore density, the resulting coefficients C of the correlation reveals a uniform picture, cf. tab. 3.5. For all foam types, the coefficient C increases with sample length, whereas the exponent varies in the range of $m = 0.63 \ldots 0.67$ for 10 ppi to 30 ppi aluminium foams, respectively. The copper foam exhibits an exponent of $m = 0.95$ to reproduce the strong rise in slope of the Nusselt number distribution with increasing sample length. The comparison of the experimental data and the second set of correlations is depicted in figs. 3.21.

Finally, an attempt is made to find a unique correlation for all foams. Since only one copper foam type is available for the measurements, which is not representative for a correlation, only the aluminium foams are considered. The correlation of type $\mathrm{Nu} = C \cdot \mathrm{Re}^m \cdot \mathrm{Pr}^{1/3}$ is applied on the

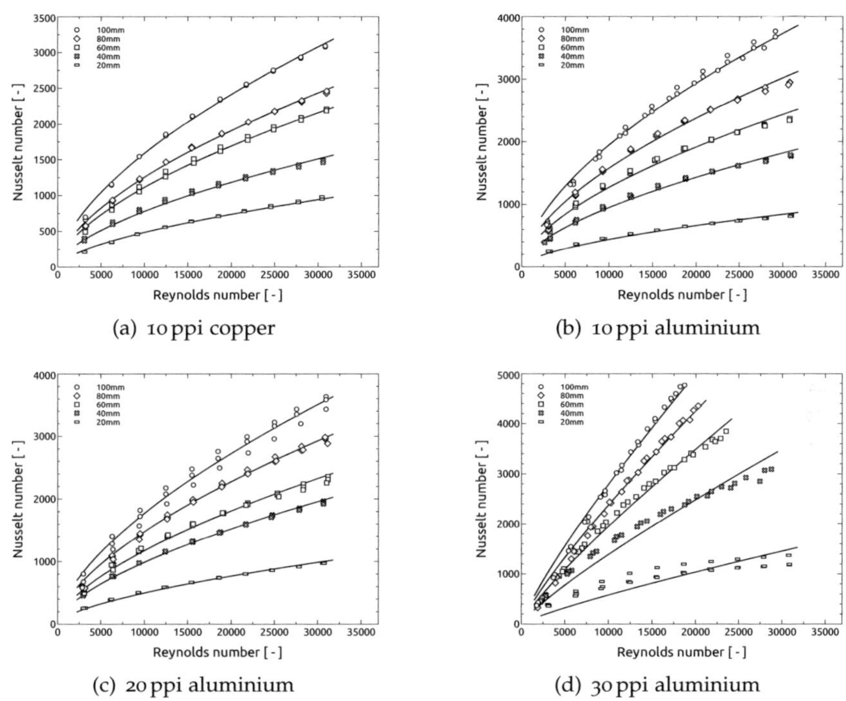

(a) 10 ppi copper

(b) 10 ppi aluminium

(c) 20 ppi aluminium

(d) 30 ppi aluminium

Figure 3.21: Nusselt number distributions of all foam types and lengths, each measured at three temperatures, 60 °C, 90 °C and 120 °C, respectively. The solid lines represent correlation curves of type $\mathrm{Nu} = C \cdot \mathrm{Re}^{m} \cdot \mathrm{Pr}^{1/3}$, for which the coefficients C are evaluated individually, whereas the exponents m are constant for each foam type; parameters are tabulated in tab. 3.5.

Nusselt number distributions of each foam type simultaneously, where the exponent m is constrained to be uniform for all lengths. Results found for the exponent $m = m(\psi)$ and the coefficients $C = C(\psi, l) = A(l)\,\psi + B(l)$, are then correlated with the relevant cumulative porosity ψ and lengths l, respectively. The symbolic structure of the correlation reads

$$\mathrm{Nu} = [A(l)\,\psi + B(l)] \cdot \mathrm{Re}^{m(\psi)} \cdot \mathrm{Pr}^{1/3}, \qquad (3.17)$$

Table 3.5: Parameters of the foam type Nusselt number correlations of type $\mathrm{Nu} = C \cdot \mathrm{Re}^{m} \cdot \mathrm{Pr}^{1/3}$. According to [179, 218] the coefficient C is fitted for each length, whereas and the exponent m is kept constant for each foam type.

foam type	Al 10 ppi	Al 20 ppi	Al 30 ppi	Cu 10 ppi
length, l [mm]		**exponent, m**		
20 ... 100	0.602	0.592	0.622	0.84
length, l [mm]		**coefficient, C**		
100	6.931	9.309	6.422	1.359
80	5.479	7.537	5.363	1.158
60	4.861	6.074	4.263	0.950
40	3.402	4.537	3.587	0.676
20	2.122	2.097	1.808	0.282

where the coefficient $A(l)$ and $B(l)$ are linearly dependent on the length l, and the exponent $m(\psi)$ is linear dependent on the porosity ψ. Least square fitting of the experimental data results in the discrete relation

$$\mathrm{Nu} = \left[\left(-0.6181\, l + 3.4856 \right) \psi + \right.$$
$$\left. + \left(0.5881\, l - 2.1673 \right) \right] \cdot \mathrm{Re}^{\left(0.5173\, \psi + 0.1748 \right)} \mathrm{Pr}^{1/3}. \qquad (3.18)$$

In comparison of the measurements with the correlation, there is good agreement for all types and lengths of the aluminium foams, cf. fig. 3.22. Differences are within a range of about $\pm$ 10 %, except for the profile for a porosity of 85.851 % at 40 mm length. The deviations are explained by the fact, that the individual samples, which are used to increment the length, evince production-related variations in porosity, cf. sec. 3.2. Again, it gets obvious that the pore density is not a sufficient measure to characterise the foams. The pore scale measures, like pore diameter or edge diameter, which result in a certain porosity, significantly influence the pressure losses as well as the heat transfer capabilities. With decreasing pore size, the specific area available for heat transfer is increasing. Furthermore, the continuous detachment and reattachment, due to the scattering of the flow in the heavily rugged pore structure, promotes the moment and energy transfer.

The resulting correlations in figs. 3.22 indicate, that with rising length the Nusselt number tends to increase with smaller porosity, for the foam

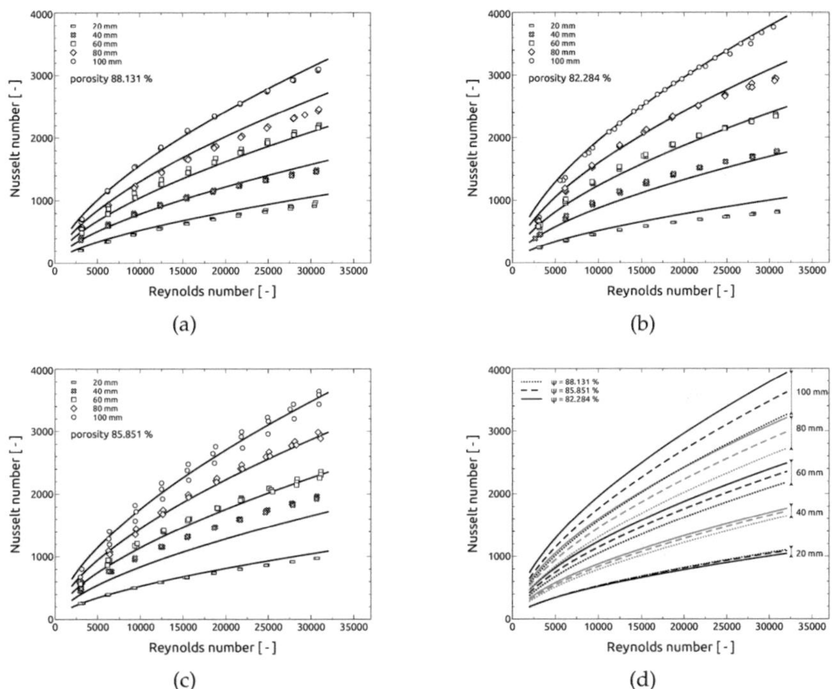

Figure 3.22: Comparison of newly developed Nusselt number correlation in comparison with the experimental data for aluminium foams of porosity (a) 88.131 %, (b) 85.851 % and (c) 82.248 % and for the lengths 20 mm, 40 mm, 60 mm, 80 mm and 100 mm, respectively. Additionally, (d) shows the comparison of the correlation for all three porosities. Each foam is measured at three temperatures, 60 °C, 90 °C and 120 °C, respectively. The solid lines represent correlation curves of type $\mathrm{Nu} = C(l, \psi) \cdot \mathrm{Re}^{m(\psi)} \cdot \mathrm{Pr}^{1/3}$, for which the coefficient $C(l, \psi)$ and exponents $m(\psi)$ are relations of length and porosity.

types under consideration, cf. fig. 3.22(d). For a length of 20 mm the curves almost coincide for different porosities, whereas the tendency is considerable for the lengths $l \geqslant 40$ mm.

3.8 Velocity profiles and massflow distributions

While the average flow velocity is determined using the massflow of the thermal massflow meter and the channel geometry, a detailed velocity distribution is obtained using the adjustable Pitot tube. This allows for qualitative and quantitative assessment of the velocity distribution at the inlet and outlet of the foam probes.

For each foam type, the minimal and maximal lengths 20 mm and 100 mm are tested for 10 operating points (discharge) and at 9 transversal locations within a massflow range of approximately 5 kg/h up to 65 kg/h. Since the massflow is controlled manually by the fan speed, there are minor differences in the massflows between the different foam types and lengths.

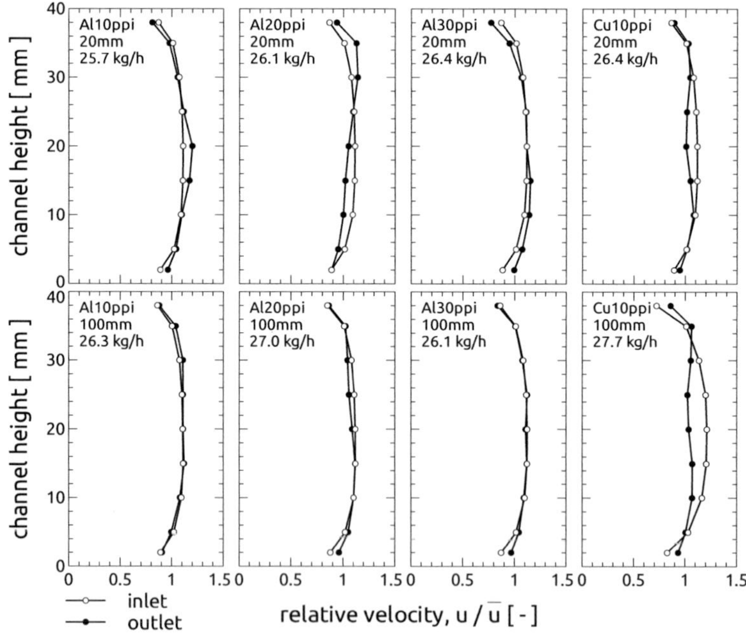

Figure 3.23: Relative velocity distributions at inlet and outlet of all foam types for 20 mm and 100 mm length, respectively.

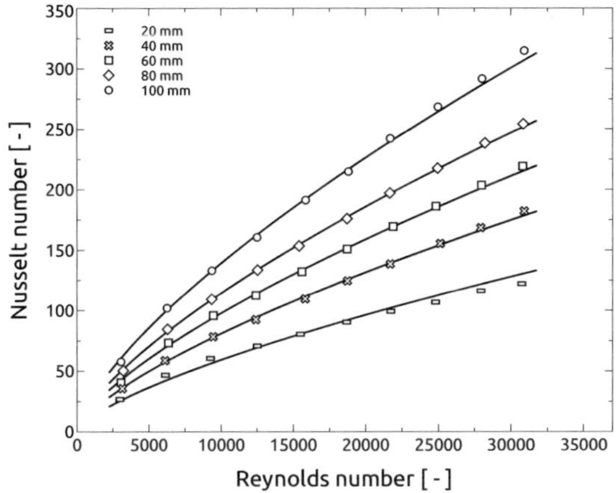

Figure 3.24: Nusselt number distributions of the reference samples of lengths 20 mm, 40 mm, 60 mm, 80 mm and 100 mm, measured at 90 °C respectively. The solid lines represent regression curves of type $Nu = C \cdot Re^{0.7} \cdot Pr^{1/3}$, for which the coefficients C are fitted individually.

The relative velocity profiles at inlet and outlet of all foam types for lengths 20 mm and 100 mm are exemplarily shown in fig. 3.23 for a massflow in the range of about 25.7 kg/h to 27.7 kg/h. In order to compare profiles at different massflows and different inlet and outlet temperatures, the relative velocity

$$\frac{u_i}{\bar{u}} = \frac{\sqrt{\frac{2 \cdot p_{\mathrm{dyn},i}}{\rho_i}} \frac{1}{n} \sum_i^n \sum_j^n \rho_j A_{\mathrm{o},i}}{\sum_i^n \rho_i A_{\mathrm{o},i} \sqrt{\frac{2 \cdot p_{\mathrm{dyn},i}}{\rho_i}}} \tag{3.19}$$

is used, which refers to the average superficial velocity and is independent from thermal conditions. For a series of transversal location $i = 1 \ldots n$, the cross section is divided into circular sections $A_{\mathrm{o},i}$. Dedicated temperature sensors and the adjustable Pitot tube are used to measure the dynamic pressure $p_{\mathrm{dyn},i}$ at these locations, whereas the density ρ_i is derived from the temperature T_i.

For all foam types with a length of 20 mm there are slight differences of inlet and outlet profile shapes. Obviously, the impact of structural and topological properties is not mixed out after a length of 20 mm. On the other hand, the inlet and outlet profiles of all aluminium foams of 100 mm length are almost congruent, whereas the copper foam of 100 mm length still shows significant differences in inlet and outlet profiles. The complete series of relative velocity profiles for all foams and all massflows is given in appendix B. For all foams of length 20 mm local disorders in the velocity profiles, due to structural and topological characteristics of the foam, evolve at small to medium discharges and are obvious up to the maximum discharge. The velocity profiles of the aluminium foams of different pore density and length 100 mm are almost evenly and homogeneous for the overall range of discharges. However, a significant kink in the outlet velocity profile at about 30 mm channel height is obvious for almost all aluminium foams of length 100 mm for the maximum discharge. This might be an indication for a turbulent wake at the outlet of the probe, which only arises at high Reynolds numbers. A very different picture is given by the copper foams of 20 mm and 100 mm length for different massflows. For both lengths, there are significant disorders and differences in the velocity profiles between inlet and outlet, that seem to evolve at small to medium discharges and keep present up to the maximum discharge. With regards to the characteristics of the copper foam, this might be due to the coarser pore structure, with larger pores and thicker edges.

As expected, the aluminium foam with highest pore density shows the smallest differences between inlet and outlet as well as the most homogeneous velocity profile. The qualitative impact of the foam structures on the characteristics of the velocity profile and flow field corresponds to the foam characteristics in terms of pore size and edge thickness. Smaller measures and filigree branch structures show less impact. In addition to the relative velocity profiles, the corresponding massflow distributions are provided in appendix C, figs. C.1 to C.6.

3.8.1 Assessment of thermal and flow characteristics

In the preceding sections pressure loss and heat transfer characteristics of different foam structures were presented. The thermal as well as the hydraulic characteristics of a heat transfer element are of vital importance for the design and qualification of heat exchangers. In order to classify and assess the performance of the foam structures for the use in heat

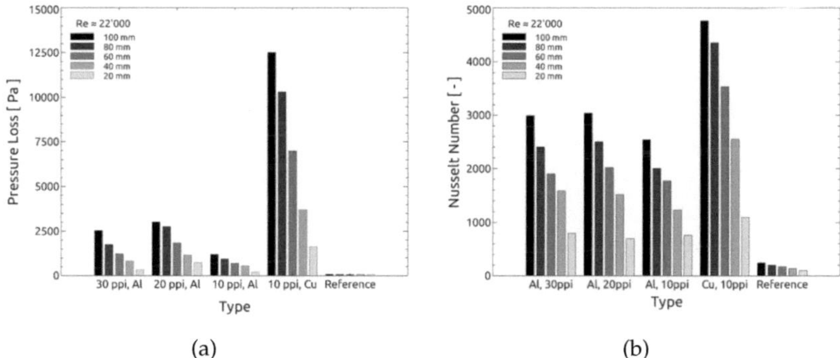

(a)   (b)

Figure 3.25: Comparison of (a) pressure loss and (b) heat transfer characteristics of different foam types and cylindrical reference samples of different lengths at a Reynolds number of approximately Re $\approx$ 22'000.

transfer applications, additional pressure loss and heat transfer measurements are carried out for cylindrical reference samples. A qualitative and quantitative comparison will show the benefits in heat transfer as well as the increased pressure losses of metal foam filled pipes in comparison to classical cylindrical channels.

The cylindrical reference samples are manufactured from seamless precision stainless steel tubes, where the inner diameter of 40 mm is equal to the cylindrical shroud diameter of the foam samples, and using the same length increments of 20 mm, 40 mm, 60 mm, 80 mm and 100 mm. Measurements of pressure loss and heat transfer characteristics are carried out equivalently to the measurements done for the foam samples above. The measured Nusselt numbers of the reference samples are depicted in fig. 3.24 for the same range of Reynolds numbers as for the foam samples. Whereas the characteristics of the Nusselt number distributions are similar to those of the foams, the magnitudes are about one order smaller. Solid line regression curves of type $Nu = C \cdot Re^{0.7} \cdot Pr^{1/3}$ are added to the diagram in order to better illustrate the characteristic of the profiles. The coefficients C are fitted individually for each length, whereas a fixed exponent of 0.7 for the Reynolds number is applied.

The pressure loss and the Nusselt number of all foam types and reference samples are shown in figs. 3.25 for all lengths. It is obvious, that using the foam samples, the heat transfer in terms of Nusselt numbers is by far increased compared to the reference samples, cf. fig. 3.25(b). On the other hand, this improvement does not come for free, since the pressure loss is increased simultaneously, cf. fig. 3.25(a).

An assessment of the foam samples relative to the reference probes is carried out by comparing the hydraulic performance in terms of the required hydraulic power

$$\dot{W}_{\mathrm{hydr}} = \Delta p \cdot \dot{V} , \tag{3.20}$$

to the so called thermal resistance, which is formulated as

$$R_{\mathrm{therm}} = \frac{1}{h\, A_{ref}} \tag{3.21}$$

according to [17]. This quantity constitutes the resistance that a body offers to a heat flux, where h is the heat transfer coefficient and A_{ref} is the characteristic heat transfer area. Both, the hydraulic and the thermal performance values are subject of minimisation with respect to efficient heat transfer applications. With decreasing thermal resistance, the easier the heat flows through the heat exchanger.

The thermal resistance is plotted against the hydraulic power for all foam types and the reference probes for all lengths in figs. 3.26. An ideal heat exchanger should exhibit small thermal resistance as well as small hydraulic power at the same time. The measured distributions in figs. 3.26 all obey the same characteristics, i.e. the thermal resistance is decreased with increasing hydraulic power, since the effect of increased convection has a positive impact on the heat transfer. Whereas the thermal resistance is of the same order of magnitude for all types of foams, the thermal resistance of the reference probes is at least one order of magnitude greater, similar to the Nusselt numbers shown in fig. 3.24. At the same time, the range of the hydraulic power of the reference samples is smaller than for the foam samples. However, foams and reference samples both share a common range, for which the foams show superior overall characteristics.

In figs. 3.27 thermal resistance vs. hydraulic power is plotted for foams and reference samples of similar lengths respectively. It is noticeable, that the different foams of the same length share almost the same characteristic distribution. Especially the aluminium foams show almost

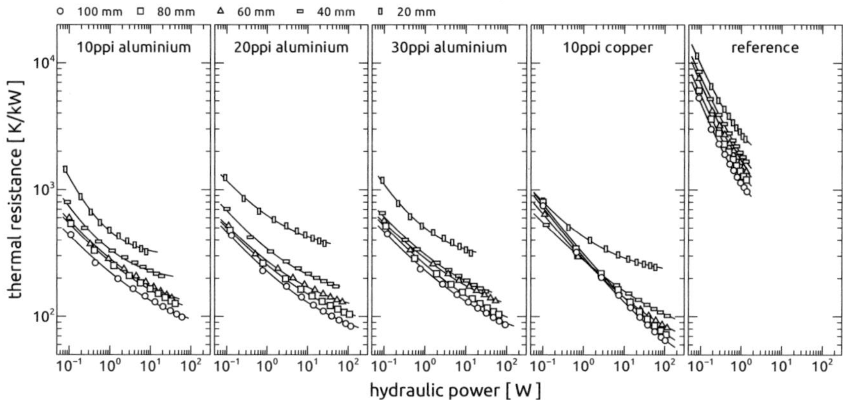

Figure 3.26: Thermal resistance vs. hydraulic power for all foam types and lengths in comparison to the tubular reference probe.

identical characteristic profiles, whereas the slope of the copper foam is slightly inclined. In general, the copper foam shows slightly better thermal performance at higher hydraulic powers. For all lengths it is obvious, that looking at the range of hydraulic power from approximately 0.1 W to 1 W, the foams show significantly less thermal resistance, and an imaginary extrapolation shows, that this holds for an even wider range of operation.

Finally, as already mentioned, the benefit in heat transfer does not come without the cost of an increased hydraulic power. In [17] foam characteristics are compared to commercially available heat exchangers, where the resistances of the foams are two times lower while requiring the same pumping power. The measurements above confirm, that the heat transfer characteristics of cellular solids, namely open cell metal foams, can be superior compared to conventional techniques, depending on the particular application, cf. [17, 94, 223].

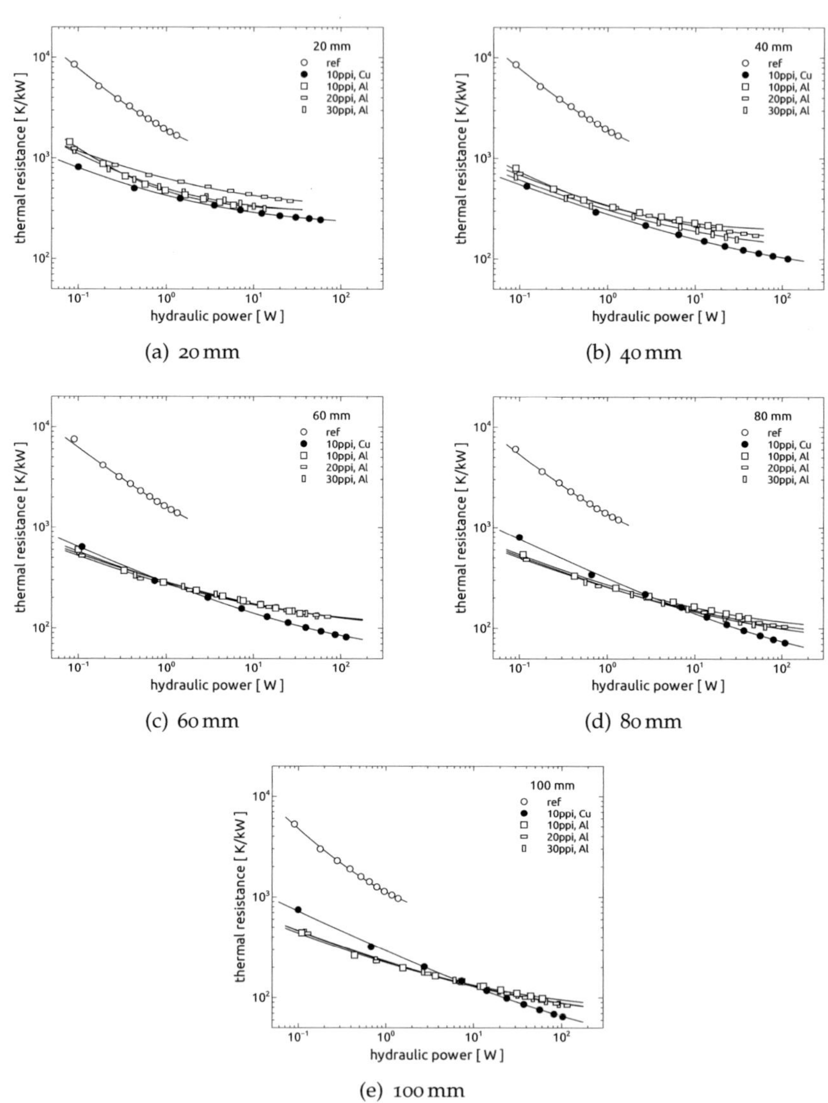

(a) 20 mm

(b) 40 mm

(c) 60 mm

(d) 80 mm

(e) 100 mm

Figure 3.27: Comparison of thermal resistance vs. hydraulic power for foam and reference samples of equal lengths 20 mm to 100 mm, (a) to (e), respectively.

Chapter 4

Diffuse interface approach

The phase-field approach enjoys a constantly increasing and wide spreading utilisation in manifold fields of applications. With its roots in solidification and crystallisation processes [58, 112, 154, 156, 185, 191] it is one of the most appropriate methods in modelling interfacial and pattern formation phenomena. Today, the areas of application include phase transition, fluid flow, crack propagation, magnetism and many more [5, 10, 14, 32, 41, 145].

Historically, the phase field approach is an easy and efficient mathematical method for solving free boundary problems. The elegance of the method lies in the fact that, compared to the classical front-tracking methods, the phase-field approach avoids the explicit treatment of phase boundaries, and the boundary conditions are implicitly applied at the interface. In the classical diffuse interface approach each of the physical entities are identified using *order parameters*, which are physical quantities exhibiting *jumps* across a phase boundary, e.g. in the Cahn-Hilliard model, the composition is an order parameter. An alternative framework, is the interpretation of the phase-field as *indicator functions* (phase-fields), wherein, each of the indicators determine the presence or absence of a given physical entity. The phase-fields vary smoothly across an artificially created interface of finite width between the physical phases. The material properties with respect to the phases are interpolated using the functionals constructed out of the phase-fields. This approach, has gained widescale utilisation owing to the ease with which the equations of motion can be derived. Consequently, the method has been applied to a wide variety of problems of phase transformations involving interfacial motion and complex geometrical evolution, also including catastrophic phase terminations.

The phase-field method models the transition between the states of different phases, where the dynamics of the phases α and β are modelled by means of the order parameters $\phi_\alpha(x, t)$ and $\phi_\beta(x, t)$, with $\phi_\alpha(x, t) = 1$

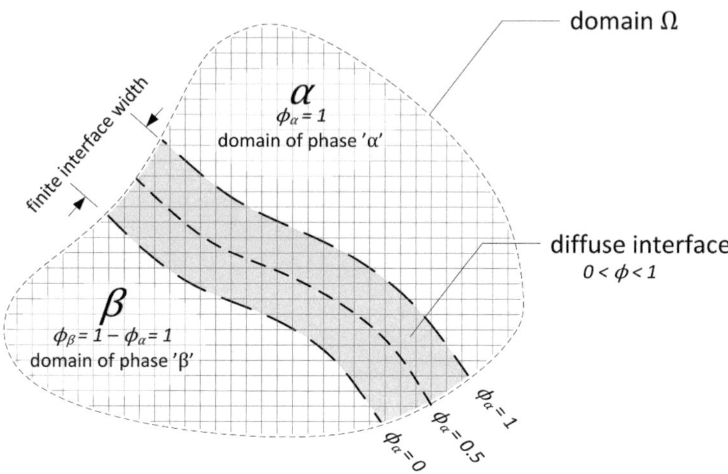

Figure 4.1: Scheme of a computational domain Ω with phases α and β. Utilising the phase field method, the interface is of finite width, and the order parameter ϕ_α changes from 1 to 0 across the diffuse interface, whereas $\phi_\beta = 1 - \phi_\alpha$. The order parameters ϕ_α and ϕ_β as well as the characteristic properties of the individual phases are constant throughout the bulk areas, respectively.

in regions where phase α is present and $\phi_\alpha(x, t) = 0$ where phase α is absent, cf. fig. 4.1. In the region of the diffuse layer surrounding phase α, ϕ_α changes continuously, differentiable and monotonously from 0 to 1. Everywhere in the simulation domain, the constraint $\phi_\alpha + \phi_\beta = 1$ must be fulfilled. In the scope of this work, we distinguish two different phases in the simulation domain, namely *solid* and *fluid* phase. In the bulk solid phase, the respective order parameter $\phi_s = 1$ whereas $\phi_f = 0$, and vice versa in the bulk fluid phase. Furthermore, the condition $\phi_s + \phi_f = 1$ holds throughout the whole numerical domain at each time.

Consider a computational domain Ω with the continuous bulk phases α and β, which are separated by a interface of finite width, cf. fig. 4.1. A equally spaced uniform Cartesian grid, which is not aligned with the interface is employed. In the bulk, the material properties are constant, whereas inside the transition region of the diffuse interface layer, either the material properties are supposed to be interpolation functions of the order parameters or a special treatment is applied, such that the physics in the bulk are maintained and asymptotically recover the physics of a sharp interface solution.

The evolution equations of the conserved variables, such as the internal energy and the mass, follow the Cahn-Hilliard approach [25] and finally result in Fick's well-known second law. Herein, without loss of generality, the energetics, the physical, thermodynamical and chemical quantities of the bulk phases, are interpolated across the diffuse interface.

The phase field method implemented in PACE3D is originally based on the model presented in [154, 156] and is under continuous development. It can simultaneously solve the mass diffusion, temperature and the front tracking in multiphase and multicomponent materials – accordingly the system variables are concentration c, temperature T and order parameter ϕ. The evolution equations are derived by the variation of the entropy functional $S(\phi, c, T)$, with respect to all system variables

$$S(\phi, c, T) = \int_\Omega \left(s(\phi, c, T) - \left(\varepsilon\, a(\phi, \nabla\phi) + \frac{1}{\varepsilon} w(\phi) \right) \right), \qquad (4.1)$$

where $s(\phi, c, T)$ is the entropy density, $a(\phi, \nabla\phi)$ is the surface gradient entropy and $w(\phi)$ is the potential. Herein, surface gradient entropy $a(\phi, \nabla\phi)$ promotes the expansion of the interface, whereas the potential $w(\phi)$ is minimal for pure phases. The formulation and balance of $a(\phi, \nabla\phi)$ and $w(\phi)$ is essential for the shape of the diffuse interface. The model parameter ε is related to the finite width λ of the diffuse interface layer. The variational derivatives of the entropy functional leads i.a. to the evolution equation of the phase field

$$\tau_k \varepsilon \frac{\partial \phi_\alpha}{\partial t} = \varepsilon \left(\nabla \cdot \frac{\partial a(\phi, \nabla\phi)}{\partial \nabla\phi_\alpha} - \frac{\partial a(\phi, \nabla\phi)}{\partial \phi_\alpha} \right) - \\ - \frac{1}{\varepsilon} \frac{\partial w(\phi)}{\partial \phi_\alpha} - \frac{1}{T} \frac{\partial f(T, \phi)}{\partial \phi_\alpha} - \Lambda, \quad (4.2)$$

where f is the bulk free energy and τ_k is the kinetic coefficient (mobility), related to the relaxation rate of the phase transition. The Lagrange multiplier Λ ensures the constraint

$$\sum_{\alpha=1}^{N} \phi_\alpha = 1, \qquad (4.3)$$

at each spatial location for all times. The most commonly used definitions of the potential $w(\phi)$ are the *multi-obstacle* and the *multi-well* potential, respectively

$$
w_{\mathrm{obst}}(\phi) = \begin{cases} \infty & \phi_\alpha = 1,\ \alpha \in [1,...,N] \\ \dfrac{16}{\pi^2} \displaystyle\sum_{\alpha<\beta} \gamma_{\alpha\beta}\, \phi_\alpha\phi_\beta \end{cases}
\tag{4.4}
$$

$$
w_{\mathrm{well}}(\phi) = 9 \sum_{\alpha<\beta} \gamma_{\alpha\beta}\, \phi_\alpha^2\phi_\beta^2 ,
\tag{4.5}
$$

where $\gamma_{\alpha\beta}$ is the surface tension between phases α and β, that scales the maxima of the potentials in $0 \leqslant \phi \leqslant 1$. The formulation yields two stable conditions at $\phi = (0,1)$, represented either by the minima of the double-well or the roots of the obstacle potential, cf. fig. 4.2. Note, that the obstacle potential is defined on the so called Gibbs-simplex $G = \{\phi \in \mathbb{R}^N : \sum_\alpha \phi_\alpha = 1,\ \ 0 \leqslant \phi_\alpha \leqslant 1\}$. The potential $w(\phi)$ is related to the order parameter by

$$
|\nabla\phi| = \frac{\partial\phi}{\partial n} = \frac{1}{\varepsilon}\sqrt{w(\phi)} .
\tag{4.6}
$$

Since this work aims for the computation of fluid flow and heat transfer in cellular solids, the essential evolution equations under consideration are the energy equation and the conservation equations of fluid mechanics, namely the Navier-Stokes equations. In the context of the phase filed method, the energy equation can be derived from the entropy functional (4.1).

According to Gauss's divergence theorem, the conservation laws for the internal energy and concentration read

$$
\begin{aligned}
\partial_t e &= \nabla \cdot j_e , \\
\partial_t c &= \nabla \cdot j_c ,
\end{aligned}
\tag{4.7}
$$

where j_e and j_e are the net energy and mass fluxes, classically given by the linear relation of the gradient of a potential, which strictly speaking only holds for $\nabla T \ll T$. For the energy equation, the flux can be defined by Fourier's law $j_e = -k\nabla T$, where the temperature T is the potential and the coefficient k is the thermal conductivity, whereas for

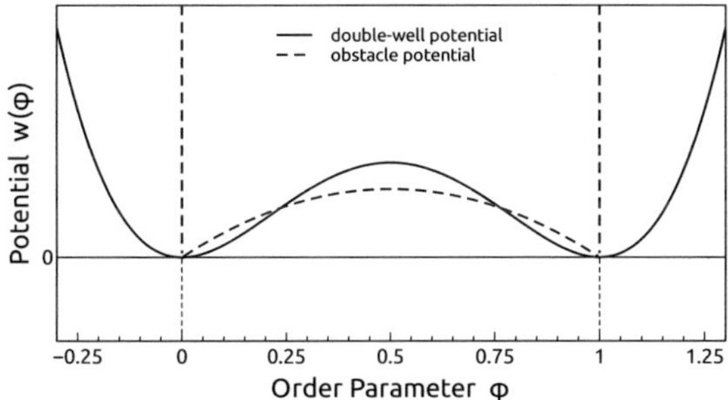

Figure 4.2: Course of the terms for the *obstacle* and *double-well* potential. Whereas the stable states are located in the minima of the double-well potential, the obstacle is defined on the Gibbs-Simplex.

the concentration Fick's law is given by $j_c = -D\nabla\rho$, with the coefficient of diffusion D.

With respect to non-equilibrium thermodynamics [164, 165] the rate of entropy generation is linearly related to the flux density. The latter is related to the thermodynamical force[1], which can be derived by the variational derivative of the entropy functional (4.1). Thus, the conservation equations of internal energy, and accordingly of concentration, as well as the evolution equation of the order parameter[2] reads

$$\partial_t e = \nabla \cdot \left(M_e^e \nabla \frac{\delta S}{\delta e} + M_e^c \nabla \frac{\delta S}{\delta c} \right) ,$$

$$\partial_t c = \nabla \cdot \left(M_c^e \nabla \frac{\delta S}{\delta e} + M_c^c \nabla \frac{\delta S}{\delta c} \right) ,$$

$$\partial_t \phi = M_\phi^\phi \frac{\delta S}{\delta \phi} ,$$

(4.8)

where M_e^i, M_c^i and M_ϕ^ϕ are the mobility coefficients of the internal energy e, the concentration c and the order parameter ϕ with respect to the system variables $i = \phi, c$ and T.

---

[1] locally defined gradients of inverse macroscopic variables

[2] since the order parameter is a non-conserved quantity, an evolution equation is formulated

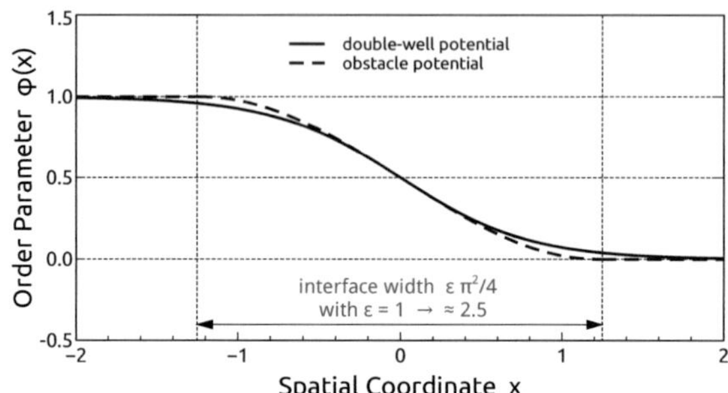

Figure 4.3: Steady state solutions of the phase field equations employing the double-well and obstacle potential, respectively. The interface width $\lambda = \varepsilon \pi^2/4$ is exact for the obstacle potential, whereas for the double-well potential it is defined as 95% of the bulk value of ϕ.

In the present work no interface dynamics and no mass diffusion is considered, i.e. $\partial_t \phi = \partial_t c = 0$. Thus, the phase field does not evolve in time, and the contributions of ϕ and c in eqns. (4.8) vanishes. The system of equations (4.8) reduces to

$$\partial_t e = \nabla \cdot \left(M_e^e \nabla \frac{\delta S}{\delta e} \right) , \qquad (4.9)$$

where the thermodynamic potential $\delta S/\delta e$ is given as $1/T$ with respect to the basic thermodynamical definition of the internal energy $e = f - TS$. Finally, Fourier's law may be written as $j = kT^2 \nabla(1/T)$, and the conservation equation of internal energy reads

$$C_V(\phi) \partial_t T = \nabla \cdot \left(k(\phi) T^2 \nabla \frac{1}{T} \right) , \qquad (4.10)$$

where the internal energy is given in terms of the heat capacity and temperature as $e = C_V(\phi)T$.

Thereby, without limiting the generality, the physical quantities of the bulk phases, $C_V(\phi)$ and $k(\phi)$ are interpolated across the diffuse interface. The derivation of an eligible interpolation formalism is essential, in order to perform quantitative simulations. Against this background, the present work contributes to the interpolation of mobility and capacity coefficients of heat conduction by means of a general diffusive transport process in chapter 6.

The conservation equations of fluid flow are not derived from the entropy functional, but coupled to the energy equation as well as to the phase-field using appropriate techniques. In the interface layer each cell is partially filled with solid and fluid according to the order parameters ϕ_s and ϕ_f, respectively. Thus, a reasonable treatment should guarantee the no-slip condition for the solid phase $u_s = 0\,\mathrm{m/s}$, an on the other hand recover the velocity profile of the bulk fluid phase. In the course of this work the lattice-Boltzmann method is employed, which asymptotically solves the Navier-Stokes equations in the limit of nearly incompressible low Mach number flows. Details of the lattice Boltzmann method as well as the coupling is presented in chapter 5.

Chapter 5

Diffuse interface fluid mechanics[1]

As a branch of continuum mechanics, fluid mechanics is concerned with the characteristics of steady and moving fluids as well as the interaction of fluids and forces. In continuum mechanics, matter – here fluid – is assumed to entirely fill the domain of interest, and to pass over the fact that it is made of individual atoms or molecules. Within the continuum hypothesis, the statistical averages of the physical properties such as density, pressure, temperature or velocity in an representative elementary volume (REV), are assumed to vary continuously within space [74].

At the macroscopic level of reality, where the density of elements is high enough, we can consider the domain of interest as a continuum. Even though individual motion is not suppressed, the large number of elements and the mutual interaction result in a superimposed collective motion. The mathematical framework is build upon the classical conservation equations of continuum mechanics, employing the specific macroscopic variables [57].

On the microscopic level, fluids can be viewed as particles or molecules that interact among themselves and with the surrounding. Within the framework of statistical mechanics, the kinetic theory establishes a rather simple mechanical picture of the particle interaction and links the molecular composition and motion to the thermodynamic entities [80, 95]. Hence, the interaction of the elements is a measure for their individuality, represented by the Knudsen-number Kn, which is defined as the ratio of the molecular free path length to characteristic physical length scale of the system. However, for systems, where $Kn \gtrsim 1$, the continuum theory no longer holds, and the methods of statistical mechanics must be applied.

---

[1]Parts of the subsequent sections are submitted for publication in *Journal of Computational Physics* [65].

In this context [121] suitably refers to the Navier-Stokes equations of continuum mechanics as a *top-down approach* and to the kinetic equations as a *bottom-up approach.*

The solution of the Navier-Stokes equations is accompanied by several numerical difficulties, the treatment of nonlinear convective terms or the solution of the Poisson equation to evaluate the pressure [161]. In lattice Boltzmann methods, the pressure is obtained by the equation of state and the nonlinear convective terms reduce to an advection type problem [219]. Compared to other common fluid dynamics applications, the lattice Boltzmann method captivates by its simplicity in terms of modelling the physical process as well as the ease of implementation. Furthermore, the lattice Boltzmann scheme is highly appropriate for parallelisation, which is important with regard to exascale computing [31, 49, 97, 99, 213].

5.1 Continuum fluid dynamics

Following the classical theory of continuum mechanics, the motion of a single phase viscous fluid in the macroscopic continuum regime, is governed by the conservation equations of mass and momentum [57, 162]

$$\partial_t \rho + \nabla \cdot (\rho \boldsymbol{u}) = 0 \tag{5.1}$$

$$\rho \left(\partial_t \boldsymbol{u} + \boldsymbol{u} \cdot \nabla \boldsymbol{u} \right) = \boldsymbol{f} - \nabla p + \nabla \cdot \mathbf{T} \tag{5.2}$$

respectively, with the density ρ, the fluid velocity $\boldsymbol{u}$, the pressure p, the body force per unit volume $\boldsymbol{f}$ and the deviatoric stress tensor $\mathbf{T}$. Employing Stokes law of friction, the latter is defined as

$$\mathsf{T}_{ij} = \mu \left(\partial_j u_i + \partial_i u_j - \frac{2}{3} \delta_{ij} \partial_l u_l \right) , \tag{5.3}$$

where μ is the dynamic viscosity and δ_{ij} is the Kronecker delta[2].

Except for a few simple, mostly academic cases, the above system of nonlinear partial differential equations can only be solved numerically [6,

---

[2] $\delta_{ij} = 1 \; \forall \, i = j$ and $\delta_{ij} = 0 \; \forall \, i \neq j$

73, 101]. The majority of technically interesting flows is turbulent, and even today the direct numerical simulation (DNS) of such flows still requires a disproportionately high effort [43]. However, the computational effort can be considerably reduced, by applying statistically averaged variables and equations[3]. Closure of this system of partial differential equations using a turbulence model, allows the application on problems of technical interest with feasible effort [171, 214].

In the course of this work comparative simulations are performed using the commercial computer-aided engineering package StarCCM, developed by CD-Adapco[4]. While the software suite was mainly developed for computational fluid dynamics (CFD) simulations, solving the Navier-Stokes equations by employing different segregated and coupled numerical algorithms, it has evolved to include additional continuum mechanics models, most notably heat transfer and solid stress models.

5.2 Kinetic theory

Whereas the most widespread methods of computational fluid dynamics (CFD) used in engineering applications today follow the classical theory of continuum mechanics by solving the Navier-Stokes equations [73, 101], methods based on the kinetic theory [80, 95], like the lattice-Boltzmann method, emerge a competent and promising alternative.

It was Bernoulli that first tried to formulate a kinetic theory in the early 17th century [11], but it took another hundred years, until the theory was proposed; amongst others by the work of Clausius, Maxwell and finally Boltzmann, cf. [15, 16, 36, 37, 139]

As part of statistical mechanics, the kinetic theory of gases is concerned with the properties and the movement of ideal gases, where it describes the gas laws from a mechanical point of view. Regarding the concept of continuum, it models a random thermal movement of a collection of particles [60, 188], uniformly with no preferred direction. Apart from collisions, the particle movement is independent and uncoupled, cf. [80, 95]. Whereas a detailed analysis of the collision and dynamics of all particles will however exceed todays available computational resources

---

[3] Reynolds Averaged Navier-Stokes equations (RANS)
[4] http://www.cd-adapco.com/ (accessed: 15/5/2014)

by far, the statistical approach of kinetic theory is the origin for various numerical methods.

5.2.1 Particle distribution function

Instead of working with individual particles, the kinetic theory considers ensembles of particles in the form of particle distribution functions adopting a statistical approach. The particle distribution function

$$f(x,\xi,t) = \frac{\mathrm{d}N}{\mathrm{d}V\mathrm{d}\xi}, \tag{5.4}$$

indicates the number of particles N in the six-dimensional phase space, spanned by the three spatial directions $\mathrm{d}V = \mathrm{d}x\,\mathrm{d}y\,\mathrm{d}z$ and the corresponding velocity components $\mathrm{d}\xi = \mathrm{d}\xi_x\,\mathrm{d}\xi_y\,\mathrm{d}\xi_z$. Thereby, the particle distribution function can be interpreted deterministically as a particle density or statistically as a probability function [95].

5.2.2 Boltzmann equation

The treatment of the particle distribution functions is in the realm of statistical mechanics, and the Boltzmann equation describes the time evolution of such a system. The differential form of the continuous Boltzmann equation

$$\frac{\mathrm{d}}{\mathrm{d}t}f(x,\xi,t) = \Omega(f), \tag{5.5}$$

relates the total derivative, the total change, of the particle distribution function $f = f(x,\xi,t)$, to the result of the collision term $\Omega(f)$ on the RHS. More explicitly one can write

$$\partial_t f + \xi \cdot \nabla_x f + \frac{1}{\rho}\mathbf{F} \cdot \nabla_\xi f = \Omega(f). \tag{5.6}$$

where still a closure is necessary by modelling the particle interaction for the collision operator on the RHS. Under the assumption of localised binary collisions in terms of the so called *Stoßzahlansatz* approximation, the collision operator takes the form of a complex integral formulation [15, 80, 95] and the Boltzmann transport equation results in a time-dependent nonlinear integro-differential equation in six-dimensional phase space, which is mathematically ambitious.

5.2.3 Maxwell-Boltzmann distribution

In the limit of thermodynamic equilibrium, the velocity distribution of the particles adopt all possible magnitudes and directions without any preference, which results in a spherical symmetry. This presumption is sufficient to derive a distribution for the thermodynamic equilibrium and to correctly calculate the microscopic and macroscopic properties [15, 139]. The so called Maxwell-Boltzmann distribution reads

$$f^{eq} = \sqrt{\left(\frac{m}{2\pi k_B T}\right)^3} \, e^{\left(-\frac{m}{2k_B T}\xi^2\right)},$$ (5.7)

where k_B is the Boltzmann constant, T is the absolute temperature, m the mass of the particle and ξ is the velocity of the particle.

5.2.4 Macroscopic quantities

In the light of statistical mechanics, the kinetic theory deals with probabilistic quantities which are characterised by different features, that are related to the moments of their probability distribution. With this in mind, the macroscopic quantities of fluid flow are determined by integrals of the product of the particle distribution function $f(x, t, \xi)$ and a function $\Psi(\xi)$ [95]. The particle density $n(x, t)$, the molecular density $\rho(x, t)$ and the velocity $v(x, t)$ for instance, are given as [95]

$$n(x, t) = \int_\xi f(x, t, \xi) \cdot d\xi$$ (5.8)

$$\rho(x, t) = m \int_\xi f(x, t, \xi) \cdot d\xi$$ (5.9)

$$v(x, t) = \frac{1}{n} \int_\xi \xi \, f(x, t, \xi) \cdot d\xi$$ (5.10)

5.2.5 BGK collision term

In order to provide a solution of the mathematically very complex Boltzmann equation (5.6), which is mainly because of the square of the particle distribution function in the collision integral [80], a simplified ansatz was suggested by Bhatnagar, Gross and Krook [13]. Herein the collision term is replaced by

$$\Omega(f) = \omega \left(f^{\text{eq}} - f \right), \tag{5.11}$$

where it is assumed, that the system relaxes towards the Maxwell-Boltzmann equilibrium distribution function f^{eq}. The relaxation time τ of the system is represented by the collision frequency $\omega = \frac{1}{\tau}$.

Despite of its simplicity, the model complies essential characteristics of the Boltzmann equation [80, 95], and it can be shown by so called Chapman-Enskog expansion [29, 115] that the BGK-Model recovers the conservation equations of continuum mechanics, namely the Euler-equations, the Navier-Stokes equations as well as the Burnett-equations.

5.3 The lattice Boltzmann method

Rather than solving the governing equations of continuum fluiddynamics directly, the lattice Boltzmann method (LBM) is based on the kinetics of statistically distributed particles in six-dimensional phase-space[5], cf. [80, 95]. It is a numerical scheme that solves the discrete Boltzmann transport equation, and it can be shown by Chapman-Enskog multiscale expansion [29], that the solution of the lattice Boltzmann equation accurately satisfies the Navier-Stokes equations for quasi-incompressible flows. Historically they are based on their predecessor the lattice gas automaton (LGA) [78, 79, 96], that suffers from statistical noise and led to the development of the LBM methods [141]. An excellent overview on the historical development of LGA and LBM as well as detailed derivations can for instance be found in [121, 194, 210, 215].

Starting from the Boltzmann evolution equation (5.6), space is discretised by a uniform, rectangular Cartesian grid, whereas the continuous velocity space is represented by a discrete set of lattice velocities e_i, linking the lattice to its nearest neighbours. A discrete velocity space $\mathbb{V}$ is introduced, i.e. the particles are allowed to follow predefined directions e_i

$$\mathbb{V} = \{e_0, \ldots, e_{q-1}\} \mid e_i \in \mathbb{R}^q, \ i = 0, \ldots, q-1, \tag{5.12}$$

---

[5]spanned by the components of spatial coordinates x and microscopic velocity ξ

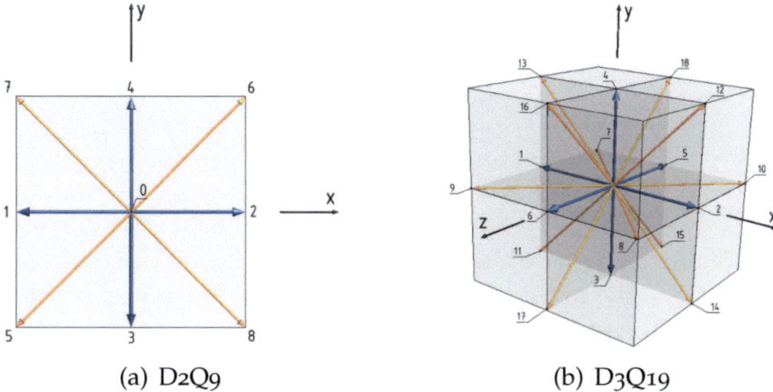

Figure 5.1: Unit lattice of the discrete velocities for (a) the two dimensional D2Q9 model and (b) the three dimensional D3Q19 model. In the unit lattice, the vectors have discrete directions as well as discrete magnitudes. For the models depicted, the vectors can have three magnitudes: 0 for particles at rest (centered at the origin of the unit lattices), 1 for particles travelling parallel to the axis (blue), $\sqrt{2}$ for particles travelling along plane diagonal (orange).

where q represents the number of discrete velocities. The employed lattice model is named DdQq, according to the number of spatial dimensions d, and the number of discrete velocities q. In PACE3D the D2Q9 model for two and the D3Q19 model for three dimensions are implemented, cf. fig. 5.1.

The lattice vectors for the D2Q9 model are given as

$$e_i = \begin{pmatrix} 0, & -1, & 1, & 0, & 0, & -1, & 1, & -1, & 1 \\ 0, & 0, & 0, & -1, & 1, & -1, & 1, & 1, & -1 \end{pmatrix}, \tag{5.13}$$

and the lattice vectors of the D3Q19 model are give as

$$e_i = \begin{pmatrix} 0 & -1 & 1 & 0 & 0 & 0 & 0 & -1 & 1 & -1 & 1 & -1 & 1 & -1 & 1 & 0 & 0 & 0 & 0 \\ 0 & 0 & 0 & -1 & 1 & 0 & 0 & 0 & 0 & 0 & 0 & -1 & 1 & 1 & -1 & -1 & 1 & -1 & 1 \\ 0 & 0 & 0 & 0 & 0 & -1 & 1 & -1 & 1 & 1 & -1 & 0 & 0 & 0 & 0 & -1 & 1 & 1 & -1 \end{pmatrix} \tag{5.14}$$

Employing the BGK collision term (5.11) in the Boltzmann equation (5.6), the discrete lattice-Boltzmann equation reads [194, 215]

$$f_i(x + e_i \Delta t, t + \Delta t) - f_i(x, t) = \Omega_{ij} \left(f_j(x, t) - f_j^{eq}(\rho, u) \right), \qquad (5.15)$$

where the f_i's are the particle distribution functions, the e_i's are the discrete velocities, x is the spatial coordinate, and Ω_{ij} is the collision matrix. Thus, the lattice Boltzmann equation is a finite difference representation of the evolution equation for the discrete particle distribution function f_i, where $\Delta x = \Delta t = 1$.

The discrete equilibrium distribution function f_i^{eq} is defined by a Taylor expansion of the Maxwell-Boltzmann equation (5.7) up to second order, and is given with respect to discrete velocity space as

$$f_i^{eq} = w_i \rho \left(1 - \frac{e_i \cdot u}{c_s^2} + \frac{(e_i \cdot u)^2}{2 c_s^4} - \frac{u^2}{2 c_s^2} \right), \qquad (5.16)$$

where $c_s = c/\sqrt{3}$ is the lattice speed of sound, c is the basic speed on the lattice with time and space step $\Delta t = \Delta x = 1$ in lattice units. From requiring that the the equilibrium distribution complies with the moments of the collision invariants mass, momentum and energy, as well as from symmetry considerations and from $\sum_i w_i = 1$, the weights are derived as [95]

$$w_i = \begin{cases} \frac{4}{9}, & i = 0 \\ \frac{1}{9}, & i = 1 \dots 4 \\ \frac{1}{36}, & i = 5 \dots 8 \end{cases} \qquad (5.17)$$

for the D2Q9 model and

$$w_i = \begin{cases} \frac{1}{3}, & i = 0 \\ \frac{1}{18}, & i = 1 \dots 6 \\ \frac{1}{36}, & i = 7 \dots 18 \end{cases} \qquad (5.18)$$

for the D3Q19 model.

5.3.1 Physical moments

The physical moments, representing the macroscopic values of density and momentum, are given by means of the discrete moments of the distribution functions according to the definitions given in sec. 5.2.4, i.e.

$$\rho = \sum_i f_i$$
$$\rho u = \sum_i e_i f_i \, . \tag{5.19}$$

5.3.2 Single relaxation time model

Applying the approximation of Bhatnagar-Gross-Krook [13], leads to the so called lattice BGK (LBGK) single-relaxation time (SRT) lattice Boltzmann model [35, 174]. Here, all moments of the distribution functions relax with the same time scale, and the simple diagonal form of the collision matrix Ω_{ij} reduces to a single relaxation parameter $1/\tau$ applied on the collision term

$$f_i(x + e_i\Delta t, t + \Delta t) - f_i(x, t) = \frac{1}{\tau}\left(f_j(x, t) - f_j^{eq}(\rho, u)\right) , \tag{5.20}$$

Herein the relaxation parameter $\omega = 1/\tau$ is related to the kinematic viscosity through

$$\nu = c_s^2\left(\tau - \Delta t\frac{1}{2}\right) = \frac{1}{3}\left(\tau - \frac{1}{2}\right) , \tag{5.21}$$

which can be derived by employing the Chapman-Enskog analysis, as shown in [95]. The lattice units are commonly normalised, i.e. $\Delta x = \Delta t = 1$, and thus it is immediately obvious, that a value of $\tau = 0.5$ is the natural lower limit of the formalism, since a negative viscosity will lead to unphysical and unstable results.

When it comes to the numerical implementation of the outlined formalism, a single time step of a typical algorithm is divided into different steps:

- The macroscopic values of ρ and u are computed as the moments of the particle distribution functions, eq. (5.19).

- During the collision step, the equilibrium distribution functions are calculated from ρ and $\boldsymbol{u}$ according to eq. (5.16), and the particle distribution functions are updated according to

$$\tilde{f}_i(\boldsymbol{x},t) = f_i(\boldsymbol{x},t) - \frac{1}{\tau}\left(f_j(\boldsymbol{x},t) - f_j^{\text{eq}}(\rho,\boldsymbol{u})\right)$$

- Finally, the new particle distribution functions $\tilde{f}_i$ are propagated towards neighbouring lattices as

$$f_i(\boldsymbol{x} + \boldsymbol{e}_i\Delta t, t + \Delta t) = \tilde{f}_i(\boldsymbol{x},t)$$

5.3.3 Multiple relaxation time model

According to the Multiple Relaxation Time (MRT) model of D'Humières [44], the numerical stability, accuracy and the physical limits of the method can be further improved, using an independently adjustable relaxation time for each kinetic moment.

With respect to the velocity-space, one can construct a moment-space for a certain DdQq model, which is formulated by the corresponding physical moments m_i, [45]. Defining a transformation matrix $\mathcal{M}$, the particle distribution functions f_i are transferred from velocity-space into the corresponding physical moments m_i in moment-space and vice versa

$$\boldsymbol{m} = \mathcal{M}\boldsymbol{f} \quad \text{or} \quad \boldsymbol{f} = \mathcal{M}^{-1}\boldsymbol{m}. \tag{5.22}$$

By applying a Gram-Schmidt orthogonalization [43], a decomposition leads to

$$\Omega_{ij} = \mathcal{M}^{-1}\hat{\Omega}_{ij}\mathcal{M} \quad \text{or} \quad \hat{\Omega}_{ij} = \mathcal{M}\,\Omega_{ij}\mathcal{M}^{-1}, \tag{5.23}$$

where the diagonal matrix $\hat{\Omega}_{ij} = \mathbb{1}\,\boldsymbol{s}$ represent the relaxation parameters s_i in moment-space, with $\mathbb{1}$ being the identity matrix. The MRT lattice Boltzmann evolution equation is written as

$$f_i(\boldsymbol{x} + \boldsymbol{e}_i\,\delta_t, t + \delta_t) = f_i(\boldsymbol{x},t) - \mathcal{M}^{-1}\,\hat{\Omega}_{ij}\left(m_i(\boldsymbol{x},t) - m_i^{(\text{eq})}(\boldsymbol{x},t)\right). \tag{5.24}$$

Collision now takes place in moment-space utilising different physically related relaxation parameters s_i. The transformations of the equilibrium moments $m_i^{(\text{eq})}$ and the values for $\mathcal{M}$ and $\hat{\Omega}_{ij}$ can be found in [45]. Due to the individual relaxation parameters, the bulk viscosity $\varsigma = \frac{2}{9}\left(\frac{1}{s_1} - \frac{1}{2}\right)$ and shear viscosity $\nu = \frac{1}{3}\left(\frac{1}{s_9} - \frac{1}{2}\right)$ are no longer coupled as in the BGK model. Thus, spurious oscillations related to low shear viscosities can be damped, and the stability of the model is improved [45, 118, 129].

Today, there is no theory to derive a perfect set of relaxation values for every application. However, values can be derived by semi empirical numerical stability analysis, studying the behaviour of the eigenvalues of a linearised collision operator [118, 129].

The transformation matrix $\mathcal{M}$ and the set of relaxation parameters for the D2Q9 model are given as

$$\mathcal{M} = \begin{pmatrix} 1 & 1 & 1 & 1 & 1 & 1 & 1 & 1 & 1 \\ -4 & 2 & -1 & 2 & -1 & 2 & -1 & 2 & -1 \\ 4 & 1 & -2 & 1 & -2 & 1 & -2 & 1 & -2 \\ 0 & -1 & -1 & -1 & 0 & 1 & 1 & 1 & 0 \\ 0 & -1 & 2 & -1 & 0 & 1 & -2 & 1 & 0 \\ 0 & 1 & 0 & -1 & -1 & -1 & 0 & 1 & 1 \\ 0 & 1 & 0 & -1 & 2 & -1 & 0 & 1 & -2 \\ 0 & 0 & 1 & 0 & -1 & 0 & 1 & 0 & -1 \\ 0 & -1 & 0 & 1 & 0 & -1 & 0 & 1 & 0 \end{pmatrix} \tag{5.25}$$

and

$$s = \begin{pmatrix} 0.0 & 1.63 & 1.14 & 0.0 & 1.92 & 0.0 & 1.92 & 1.99 & 1.99 \end{pmatrix} \tag{5.26}$$

whereas the inverse transformation matrices $\mathcal{M}^{-1}$ and the values for the D3Q19 model are listed in appendix D. The numerical implementation is similar to the one outlined for the SRT model, except that the collision is carried out in the moment space.

5.4 Linkage to Navier-Stokes equation – Chapman-Enskog expansion

The multiscale singular perturbation analysis known as Chapman-Enskog expansion allows to derive the Navier-Stokes equations from the Boltzmann equation by means of an asymptotic analysis [121, 209, 215]. The following description is mainly inspired by [209] and is presented here for completeness and comprehension. Starting from the discrete Boltzmann equation with a general collision term $\Omega(x,t)$, a Taylor expansion for a function of two variables for $f_i(x + c_i, t + 1)$ leads to

$$\Omega_i(x,t) \approx (\partial_t + \nabla \cdot c_i)\, f_i + \frac{1}{2}(\partial_t^2 + 2\partial_t \nabla \cdot c_i + \nabla\nabla : c_i c_i)\, f_i + \mathcal{O}(3)$$

$$(5.27)$$

Perturbation theory provides the technique to separate the function $f_i(x,t)$ in two timescales t_1 and t_2 for fast and slow phenomena (advection and diffusion), respectively. The time derivative is expanded as

$$\partial_t = \epsilon \partial_{t_1} + \epsilon^2 \partial_{t_2} + \mathcal{O}(\epsilon^3)\,, \tag{5.28}$$

where ϵ is the perturbation parameter. Similarly, the spatial derivative is expanded for one term

$$\nabla = \epsilon \nabla_1 + \mathcal{O}(\epsilon^2)\,. \tag{5.29}$$

Finally, the particle distribution function as well as the collision term are expanded, starting at the zeroth order terms $f_i^{(0)}$ and $\Omega_i^{(0)}$ as

$$f_i = f_i^{(0)} + \epsilon f_i^{(1)} + \mathcal{O}(\epsilon^2) \tag{5.30}$$

$$\Omega_i = \Omega_i^{(0)} + \epsilon \Omega_i^{(1)} + \epsilon^2 \Omega_i^{(2)} + \mathcal{O}(\epsilon^3) \tag{5.31}$$

Summarising the expansions above and substitute them in the Taylor series (5.27) reads

$$\Omega_i^{(0)} + \epsilon \Omega_i^{(1)} + \epsilon^2 \Omega_i^{(2)} + \mathcal{O}(\epsilon^3) =$$

$$\left(\epsilon \partial_{t_1} + \epsilon^2 \partial_{t_2} + \epsilon \nabla_1 \cdot c_i + \epsilon^2 \frac{1}{2} \partial_{t_1}^2 + \epsilon^2 \partial_{t_1} \nabla_1 \cdot c_i + \epsilon^2 \frac{1}{2} \nabla_1 \nabla_1 : c_i c_i \right)$$

$$\left(f_i^{(0)} + \epsilon f_i^{(1)} \right) + \mathcal{O}(\epsilon^3)\,. \tag{5.32}$$

Sorting and rearranging with respect to the smallness parameter ϵ one gets

$$\Omega_i^{(0)} + \epsilon\Omega_i^{(1)} + \epsilon^2\Omega_i^{(2)} + \mathcal{O}(\epsilon^3) = \epsilon\left[(\partial_{t_1} + \nabla_1 \cdot c_i)f_i^{(0)}\right] +$$
$$+ \epsilon^2\left[(\partial_{t_1} + \nabla_1 \cdot c_i)f_i^{(1)} +\right.$$
$$\left. + \left(\partial_{t_2} + \frac{1}{2}\partial_{t_1}^2 + \partial_{t_1}\nabla_1 \cdot c_i + \frac{1}{2}\nabla_1\nabla_1 : c_ic_i\right)f_i^{(0)}\right] +$$
$$+ \mathcal{O}(\epsilon^3). \quad (5.33)$$

Looking at the terms of different order in ϵ separately by means of a coefficient comparison, it directly follows that $\Omega_i^{(0)} = 0$, and for the remaining terms

$$\Omega_i^{(1)} = (\partial_{t_1} + \nabla_1 \cdot c_i)f_i^{(0)} \qquad\qquad (5.34)$$

$$\Omega_i^{(2)} = (\partial_{t_1} + \nabla_1 \cdot c_i)f_i^{(1)} +$$
$$+ \left(\partial_{t_2} + \frac{1}{2}\partial_{t_1}^2 + \partial_{t_1}\nabla_1 \cdot c_i + \frac{1}{2}\nabla_1\nabla_1 : c_ic_i\right)f_i^{(0)} \quad (5.35)$$

Conserved quantities and BGK collision term

In order to proceed with the linkage between the lattice Boltzmann equation and the Navier-Stokes equations, we revise the definition of the conserved macroscopic quantities as moments of the particle distribution function here

$$\rho = \sum_i f_i = \sum_i f_i^{(0)} \qquad\qquad (5.36)$$

$$\rho u = \sum_i c_i f_i = \sum_i c_i f_i^{(0)} \qquad\qquad (5.37)$$

$$\Pi = \sum_i Q_i f_i \qquad\quad \text{with } Q_i = c_i c_i - c_s^2 \qquad (5.38)$$

Furthermore, the BGK collision operator is expressed by expansion of the particle distribution function around $f_i^{(0)}$ as

$$\Omega_i = -\frac{1}{\tau}\left(f_i - f_i^{(0)}\right) = -\frac{1}{\tau}\left(\epsilon f_i^{(1)} + \epsilon^2 f_i^{(0)} + \mathcal{O}(\epsilon^3)\right) \qquad (5.39)$$

Mass and Momentum are preserved by the collision operator, i.e. $\sum_i \Omega_i = \sum_i c_i \Omega_i = 0$, which further leads to

$$\sum_i \Omega_i = -\frac{1}{\tau}\sum_i f_i^{(k)} = 0 \qquad \forall\, k > 0 \qquad (5.40)$$

$$\sum_i c_i \Omega_i = -\frac{1}{\tau}\sum_i c_i f_i^{(k)} = 0 \qquad \forall\, k > 0 \qquad (5.41)$$

i.e. for the BGK operator, the zeroth and first order moments of the non-equilibrium distribution will vanish.

$\mathcal{O}(0)$ moments of the Chapman-Enskog expansion

Formulating the zeroth order moment for the terms of $\mathcal{O}(\epsilon)$ from the Chapman-Enskog expansion, eq. (5.34), reads

$$\sum_i \Omega_i^{(1)} = \sum_i \partial_{t_1} f_i^{(0)} + \sum_i \nabla_1 \cdot c_i f_i^{(0)} = 0. \qquad (5.42)$$

With respect to the results for the conserved quantities, eq. (5.36) and (5.37), it follows

$$\sum_i \partial_{t_1} f_i^{(0)} = \partial_{t_1} \sum_i f_i^{(0)} \qquad \text{with} \quad \sum_i f_i^{(0)} = \rho \qquad (5.43)$$

$$\sum_i \nabla_1 c_i f_i^{(0)} = \nabla_1 \sum_i c_i f_i^{(0)} \qquad \text{with} \quad \sum_i c_i f_i^{(0)} = \rho u. \qquad (5.44)$$

Putting everything together, the continuity equation is recovered for timescale t_1

$$\partial_{t_1}\rho + \nabla_1 \cdot \rho u = 0. \qquad (5.45)$$

In the same way the zeroth order moment for the terms of $\mathcal{O}(\epsilon^2)$ from the Chapman-Enskog expansion, eq. (5.35), reads

$$\sum_i \Omega_i^{(2)} = \partial_{t_1} \sum_i f_i^{(1)} + \nabla_1 \cdot \sum_i c_i f_i^{(1)} + \partial_{t_2} \sum_i f_i^{(0)} +$$

$$\frac{1}{2}\partial_{t_1}^2 \sum_i f_i^{(0)} + \partial_{t_1} \nabla_1 \cdot \sum_i c_i f_i^{(0)} + \frac{1}{2}\nabla_1\nabla_1 : \sum_i c_i c_i f_i^{(0)} = 0. \quad (5.46)$$

Applying eq. (5.40) we can drop the first two terms on the right hand side. Additionally applying the definitions of the conserved quantities, eqns. (5.36) to (5.38), this simplifies to

$$\partial_{t_2}\rho + \frac{1}{2}\partial_{t_1}^2\rho + \partial_{t_1}\nabla_1 \cdot \rho u + \frac{1}{2}\nabla_1\nabla_1 : \sum_i \left(\mathbf{Q}_i + c_s^2 \mathbb{1}\right) f_i^{(0)} = 0, \quad (5.47)$$

and with $\nabla_1\nabla_1 : \mathbb{1} = \nabla_1^2$ it finally follows

$$\partial_{t_2}\rho + \partial_{t_1}\nabla_1 \cdot \rho u + \frac{1}{2}\left(\partial_{t_1}^2\rho + \nabla_1\nabla_1 : \mathbf{\Pi}^{(0)} + c_s^2\nabla_1^2\rho\right) = 0. \quad (5.48)$$

$\mathcal{O}(1)$ moments of the Chapman-Enskog expansion

The first order moment for the terms of $\mathcal{O}(\epsilon)$ from the Chapman-Enskog expansion, eq. (5.34), reads

$$\sum_i c_i \Omega_i^{(1)} = \partial_{t_1} \sum_i c_i f_i^{(0)} + \nabla_1 \cdot \sum_i c_i c_i f_i^{(0)} = 0, \quad (5.49)$$

which results in

$$\partial_{t_1}\rho u + \nabla_1 \cdot \mathbf{\Pi}^{(0)} + c_s^2\nabla_1\rho = 0. \quad (5.50)$$

Similarly, the first order moment of eq. (5.35) is given as

$$\sum_i c_i \Omega_i^{(2)} = \partial_{t_1} \sum_i c_i f_i^{(1)} + \nabla_1 \cdot \sum_i c_i c_i f_i^{(1)} + \partial_{t_2} \sum_i c_i f_i^{(0)} +$$

$$\frac{1}{2}\partial_{t_1}^2 \sum_i c_i f_i^{(0)} + \partial_{t_1}\nabla_1 \cdot \sum_i c_i c_i f_i^{(0)} + \frac{1}{2}\nabla_1\nabla_1 : \sum_i c_i c_i c_i f_i^{(0)} = 0, \quad (5.51)$$

which leads to

$$
\nabla_1 \cdot \mathbf{\Pi}^{(1)} + \partial_{t_2}\rho u + \frac{1}{2}\partial_{t_1}^2 \rho u + \partial_{t_1}\nabla_1 \cdot \mathbf{\Pi}^{(0)} + c_s^2 \partial_{t_1}\nabla_1 \rho +
$$
$$
+ \frac{1}{2}\nabla_1\nabla_1 : \mathbf{R}^{(0)} = 0, \quad (5.52)
$$

where $\mathbf{R}^{(0)}$ is a rank 2 tensor $(R^{(0)}_{\alpha\beta\gamma} = \sum_i c_{i\alpha}c_{i\beta}c_{i\gamma}f_i^{(0)})$ which is be further investigated in the course of an Chapman-Enskog analysis for the momentum equation, cf. [209].

Deriving the continuity equation

The expansion as well as the zeroth and first order moments derived so far can now be combined. Taking one half of the time derivative in t_1 of eq. (5.45) and substracting it from eq. (5.48) gives

$$
\partial_{t_2}\rho + \frac{1}{2}\left(\partial_{t_1}\nabla_1 \cdot \rho u + \nabla_1\nabla_1 : \mathbf{\Pi}^{(0)} + c_s^2 \nabla_1^2 \rho \right) = 0. \quad (5.53)
$$

On closer inspection it gets obvious, that the expression in brackets is identical to the divergence of eq. (5.50), which means that

$$
\partial_{t_2}\rho = 0. \quad (5.54)
$$

While t_1 is the time scale of fast phenomena, the result for the timescale t_2 states, that the particle exchange due to diffusion takes place without a redistribution of mass, which obeys to advective processes. Since $\partial_t \rho = \epsilon \partial_{t_1}\rho + \epsilon^2 \partial_{t_2}\rho$ and $\nabla = \epsilon \nabla_1$, it follows that

$$
\partial_t \rho + \nabla \cdot \rho u = 0, \quad (5.55)
$$

which is the continuity equation for compressible fluids. The derivation of the momentum equations is done by the same analysis but is much more involved. Therefore the interested reader may refer [95, 121, 209, 215]. At this point it is noteworthy, that the simple collision and streaming scheme presented above, provides a feasible and satisfying technique to render the Navier-Stokes equations up to second order in space and time [106].

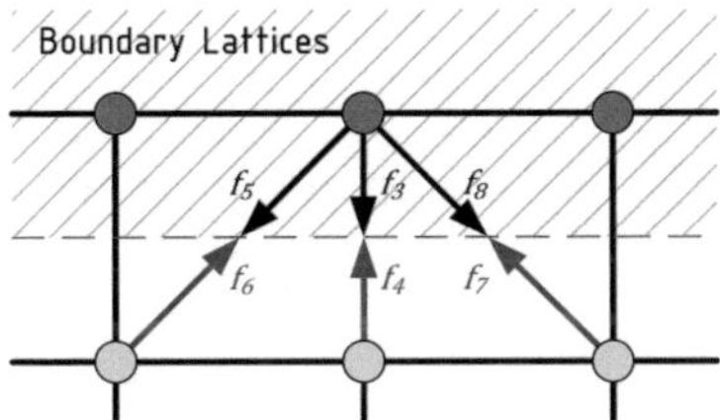

Figure 5.2: Illustration of the *bounce back* mechanism for a two-dimensional D2Q9 lattice. Incoming particle distribution functions (f_4, f_6 and f_7) are reflected (f_3, f_5 and f_8) applying a rather mechanical concept. Herein, the no-slip wall is located halfway between the *fluid* and *boundary* lattices.

5.5 Boundary conditions

One of the advantages of the lattice Boltzmann method is the easy and seamless implementation of basic fluiddynamic boundary conditions. On the other hand, since boundary conditions will have to be formulated in terms of the distribution functions rather than in terms of physical quantities, the translation of macroscopic to microscopic information can impose considerable difficulties, demanding for more involved schemes, cf. [87, 98, 224].

In the following, only a brief description of a few simply boundary treatments which were used in this work is presented. However, there are by far more boundary condition schemes available in the open literature, which are more sophisticated [21, 87, 98, 131, 186, 224], but a discussion of those would go beyond the scope of this work.

Bounce Back: The so called *bounce back* boundary condition is realised by applying a rather simple mechanism associated with a mechanical viewpoint. To render the hydraulic no-slip boundary condition, particle distribution functions at the walls are simply reflected, cf. fig. 5.2. The standard bounce-back scheme is explained and described at length in the open literature, cf. [81, 150, 196]. Problems may arise when the physical wall is not aligned with the lattices, which requires for more involved schemes or certain interpolation techniques, cf. [21, 34, 87, 91, 160].

Equilibrium Boundary Condition: Employing the so called *equilibrium boundary condition*, the particle distribution functions f_i of the concerned lattice are simply

$$f_i(x, t) = f_i^{\text{eq}}(u_b), \tag{5.56}$$

i.e. subject of equilibrium distribution according to the velocity u_b. In [151] it is shown, that for moderate Reynolds numbers in the weak non-equilibrium regime, these kind of boundary conditions provide a reasonable option.

Periodic Boundary Conditions: Often, when it comes to represent periodic flow conditions, the problem setup might be substantially reduced, especially in terms of computational resources and simulation time. Periodic boundaries always appear in pairs, i.e. flow that leaves on one boundary of a domain is supposed to enter at the adjacent boundary and vice versa, for instance. Accordingly, the quantities of the corresponding boundaries are exchanged, whereas employing the lattice Boltzmann method distribution functions are exchanged rather than physical quantities.

Extrapolation Boundary Condition: At outlet or opening boundaries, where the values or conditions of the physical quantities are not known at first, the most appropriate assumption one could make, is to extrapolate the interior values onto the domain boundary. However, this might lead to stability problems, especially for high gradients or recirculating flows at the concerning boundary. Stability might be slightly improved by a weighting of the distribution function extrapolation in terms of pressure or density.

5.6 Modelling of body forces

So far body forces are omitted for simplicity. However, in order to simulate gravity or buoyancy driven flows, and for the modelling of the diffuse interface coupling, a suitable treatment is required. There are different models proposed and discussed for the incorporation of body forces into lattice Boltzmann method, cf. [23, 92, 117, 127, 136, 150].

Following [142, 143] and with regards to the diffuse interface modelling according to [10] which will be described in sec. 5.8, we employ the exact difference method proposed by [117]. The velocity change due to the action of the force $\Delta u = F\Delta t/\rho$ is used to model the force contribution

$$\Delta f_i = f_i^{(\mathrm{eq})}(\rho, u + \Delta u) - f_i^{(\mathrm{eq})}(\rho, u)\,, \tag{5.57}$$

that is added to the right hand side of eq. (5.15).

5.7 Smagorinsky subgrid-scale turbulence model

To account for the effects of spatially unresolved turbulent effects we employ the subgrid-scale model of Smagorinsky [104, 171, 172, 180, 187], where the eddy-viscosity ν_t is related to

$$\nu_t = (C_s\Delta)^2\,|\mathcal{S}|\,, \tag{5.58}$$

with the Smagorinsky constant C_s, the filter size Δ, and the magnitude $|\mathcal{S}|$ of the strain rate tensor $\mathcal{S}_{ij} = (\partial_i u_j + \partial_j u_i)$

$$|\mathcal{S}| = \sqrt{2\,\mathcal{S}_{ij}\mathcal{S}_{ij}}\,. \tag{5.59}$$

In LBE the strain rate tensor $\mathcal{S}_{ij}$ can be calculated directly from the second-order moment of the non-equilibrium distribution function [104, 220] as

$$\mathcal{S}_{ij} = \frac{3}{2\rho\tau_t\Delta t}\Pi_{ij}^{(\mathrm{neq})}\,, \tag{5.60}$$

with the non-equilibrium momentum tensor

$$\Pi_{ij}^{(\mathrm{neq})} = \sum_\alpha e_{\alpha i}e_{\alpha j}(f_\alpha - f_\alpha^{(\mathrm{eq})})\,. \tag{5.61}$$

From this, the eddy-viscosity can be expressed as

$$\nu_t = \frac{3}{2\rho\tau_t\Delta t}(C_s\Delta)^2\sqrt{\Pi_{ij}^{(\mathrm{neq})}\Pi_{ij}^{(\mathrm{neq})}}\,. \tag{5.62}$$

Finally, we use the effective viscosity $\nu^* = \nu + \nu_t$ to get an effective relaxation time

$$\tau^* = \frac{3\nu^* + 1}{2}\,. \tag{5.63}$$

Using the effective relaxation time in eq. 5.20, we model a filtered form of the LBE in terms of a large-eddy simulation [104].

5.8 Diffuse interface modelling

With respect to the diffuse interface region of the phase field method, a special treatment of the flow in the phase transition region of fluid (f) and solid (s) phases is essential, cf. chapter 4. Inside the diffuse interface layer each cell is partially filled with solid and fluid according to the magnitude of the order parameters ϕ_s and ϕ_l, respectively. An appropriate modelling approach should recover the no-slip condition for the solid phase $u_s = 0\,\mathrm{m/s}$ as well as the velocity profile of the bulk fluid phase. In the following two different models are presented and validations are carried out in the subsequent sections.

5.8.1 Diffuse bounce back

A notably simple method is proposed in [148], where the streaming step is modified, such that depending on the order parameter ϕ only a fraction of the distribution function is transmitted. The modified evolution equation for the SRT lattice Boltzmann method reads

$$f_i(x + e_i\,\delta_t, t + \delta_t) = \tilde{f}_i(x,t) - r(\phi)\,\Omega_{ij}\left(f_i - f_i^{(\mathrm{eq})}\right), \qquad (5.64)$$

where the reflection is carried out by means of a diffuse bounce back mechanism

$$\tilde{f}_i(x,t) = r(\phi)f_i(x,t) + (1 - r(\phi))f_{\bar{i}}(x,t). \qquad (5.65)$$

Herein $f_{\bar{i}}(x,t)$ denotes particle distribution functions bounced back in opposite direction with respect to i. The reflectivity function $r(\phi)$ is modelled according to [148] and with respect to the steady state solution of the diffuse interface profile as

$$r(\phi) = \frac{1}{2}\left(1 - \tanh\left(\frac{d_w}{\lambda_f}\right)\right), \qquad (5.66)$$

where $d_w = \lambda\,\mathrm{atanh}(0.5 - \phi_f)/\delta x$ is the distance relative to the interface ($\phi = 0.5$) and $\lambda_f = v/c_s$ is the molecular mean free path. Furthermore, ϕ_f is the liquid fraction and $\lambda = \epsilon\pi^2/4$ is the interface width. Since the reflectivity is given as a function of the diffuse phase-field variables, a smooth transition at the phase boundary (fluid-solid) is guaranteed. We will reference this model as *reflectivity* model the following.

5.8.2 Dissipative interfacial stress term

Following the methodology of [10], the effect of the diffuse interface region can also be modelled by an additional dissipative forcing term for the partially filled regions, cf. [142, 143, 205]. The same approach is recently applied by [144] for the coupled phase field and lattice Boltzmann simulation of dendritic solidification including the effects of liquid motion. The phase-averaged velocity $\bar{u}$ inside the diffuse interface region, with respect to $u_s = 0$, is written as the volume averaged velocity $\bar{u} = u_s\phi_s + u_l\phi_l = u_l\phi_l$. According to Beckermann et al. [10], the fluid is considered incompressible, which corresponds well with the low Mach number assumption of the lattice Boltzmann method. Hence, the conservation equations of fluid mechanics can be formulated as [10]

$$\nabla \cdot (\phi_f u) = 0$$

$$\partial_t(\phi_f \rho u) + \nabla \cdot \left(\phi_f \rho u u^T\right) = -\phi_l \nabla p + \nabla \cdot (\mu \nabla (\phi_f u)) - \langle [\tau]^f \cdot \nabla X_f \rangle_V$$

$$(5.67)$$

where the last term on the right hand side of the second equation $M_f^d = \langle [\tau]^f \cdot \nabla X_f \rangle$ is derived from the interfacial momentum source. The term is named *dissipative interfacial stress term*, and accounts for additional dissipation viscous stress inside the interface layer. Herein τ corresponds to the viscous stress tensor.

In accordance with flow inside porous media [113, 157, 189] the dissipative interfacial stress term is added to the liquid phase and is modelled according to Darcy's law for porous media flow as

$$M_f^d = h\mu_f(1 - \phi_f)\frac{u_f}{\delta}|\nabla\phi|, \qquad (5.68)$$

with a dimensionless friction coefficient h, a model parameter relative to interface width δ, the liquid dynamic viscosity μ_f and the solid fraction ϕ_s. Note, that δ is a parameter related to the interface thickness, which is given as approximately 6δ when using the double well potential formulation $\phi(n) = \frac{1}{2}(1 + \tanh(n/2\delta))$ given in [10]. Since the double well will give $\phi = 0$ or 1 only as $x \to \pm\infty$ respectively, the interface thickness is defined as 6δ for the range of $0.05 \leqslant \phi \leqslant 0.95$, whereas the interface width of the obstacle potential eq. (4.4) is defined as $\lambda = \pi^2\varepsilon/4$

for $0 \leqslant \phi \leqslant 1$. Comparing the interface widths $6\delta \approx \lambda$ and ignoring the difference between double-well and obstacle potential gives

$$\delta = \pi^2 \varepsilon / 24 \,. \tag{5.69}$$

For the binary $\alpha|\beta$ interface at equilibrium $\partial_t \phi = 0$ the balance of the variations of surface gradient energy and potential

$$\varepsilon \left(\frac{\partial a(\phi, \nabla\phi)}{\partial \phi} - \nabla \cdot \frac{\partial a(\phi, \nabla\phi)}{\partial \nabla\phi} \right) = \frac{1}{\varepsilon} \frac{\partial w(\phi)}{\partial \phi} \,. \tag{5.70}$$

The surface gradient energy

$$a(\phi, \nabla\phi) = \sum_{\alpha < \beta} \gamma_{\alpha\beta} (a_{\alpha\beta}(\phi, \nabla\phi))^2 |q_{\alpha\beta}|^2 \,, \tag{5.71}$$

can be simplified, since $a_{\alpha\beta} \equiv 1$ in case of isotropy and $q_{\alpha\beta} = -\nabla\phi$ for a binary $\alpha|\beta$ interface, to

$$a(\phi, \nabla\phi) = \gamma_{\alpha\beta} |\nabla\phi|^2 \,, \tag{5.72}$$

Using this in eq. (5.70) for the one-dimensional case, leads to

$$2\varepsilon \gamma_{\alpha\beta} \nabla^2 \phi = \frac{1}{\varepsilon} \frac{\partial w(\phi)}{\partial \phi} \,, \tag{5.73}$$

multiplying with $\nabla\phi$ and integrating with respect to ϕ

$$\varepsilon \gamma_{\alpha\beta} |\nabla\phi|^2 = \varepsilon a(\phi, \nabla\phi) = \frac{1}{\varepsilon} w(\phi) \,, \tag{5.74}$$

and using the obstacle potential (4.4) gives

$$|\nabla\phi| = \frac{4}{\pi\varepsilon} \sqrt{\phi(1-\phi)} \,. \tag{5.75}$$

Putting things together, using eqns. (5.69) and (5.75) in eq. (5.68) and $u_l = u/\phi_l$, the dissipative interfacial stress term reads

$$M_f^d = \frac{96}{\pi^3 \epsilon^2} h \mu_f u \frac{\phi_s}{\phi_f} \sqrt{\phi_s \phi_f} \,. \tag{5.76}$$

All terms are now defined except the friction coefficient h, which is derived from a plain Poiseuille flow. Hence, the momentum equation (5.67) simplifies to

$$\mu_f \frac{\partial^2(\phi_f u)}{\partial x^2} - \frac{96}{\pi^3 \epsilon^2} h\, \mu_f\, u\, \frac{\phi_s}{\phi_f} \sqrt{\phi_s\, \phi_f} = \phi_l \frac{\partial p}{\partial y}. \tag{5.77}$$

Since there is no analytical solution for eq. (5.77), it is normalised with respect to $X = x/\Delta x$ and solved numerically according to [193] using Mathematica. According to [10] the ansatz $u(X) = \exp(\sqrt{h}X)$ is used for the velocity. Finally, the friction coefficient is evaluated as $h = 3.285$. In the context of the lattice Boltzmann method, the additional dissipative interfacial stress term is added by means of the body force modelling mentioned in sec. 5.6 using the formalism of [117], and the model is named *forcing model* within the subsequent sections.

5.9 Parametrisation

As mentioned at different occasions above, the lattice Boltzmann formalism implies that the dimensionless set of space and time unit equals unity, $\Delta x = \Delta t = 1$. However, the simulation strives to reflect actual, real world physics. With regards to the laws of similitude (Buckingham Π-theorem) [217], flows are equivalent if they achieve geometric-, kinematic- and dynamic similarity. The solution of the incompressible Navier-Stokes equation is mainly characterised by the Reynolds number.

With respect to the prerequisite of small Mach number regime[6] for the lattice Boltzmann method, the flow can be considered as effectively incompressible. Thus, the Reynolds number is held to be the same for both the physical and the lattice units, in order to ensure dynamic similitude between the real world and simulation.

Starting from the physical system, the characteristic Reynolds number is given by the characteristic length scale l_0, velocity u_0 and viscosity ν_0 as

$$\mathrm{Re} = \frac{l_0 \cdot u_0}{\nu_0}. \tag{5.78}$$

---

[6]The equilibrium distribution only holds for small velocities or low Mach numbers [95, 215].

Note, that in an intermediate dimensionless system (d) on the way from physical to lattice units, the Reynolds number is given as the reciprocal of the dimensionless viscosity $\mathrm{Re} = 1/\nu_d$, since $\Delta x = \Delta t = 1$.

Now an appropriate choice on the spatial resolution N_x and the velocity $u_{0,(\mathrm{lb})}$ in lattice units (lb) must be made to fix the discrete cell spacing δ_x and the discrete time step δ_t as

$$\delta_x = \frac{l_0}{N_x} \tag{5.79}$$

$$\delta_t = \frac{u_{0,(\mathrm{lb})}}{u_0} \cdot \delta_x \tag{5.80}$$

The decision taking and the constraints for the choice of N_x and $u_{(\mathrm{lb})}$ will be more clear after the whole picture of the parametrisation is drawn.

Typically, the velocity in lattice units is the one applied as boundary or initial condition, otherwise an arbitrary physical velocity $u_{i,(\mathrm{p})}$ is scaled to lattice units as

$$u_{i,(\mathrm{lb})} = u_{i,(\mathrm{p})} \cdot \frac{u_{0,\mathrm{lb}}}{u_0} , \tag{5.81}$$

whereas the viscosity in lattice units is given as

$$\nu_{(\mathrm{lb})} = \nu_0 \cdot \frac{\delta_t}{\delta_x^2} . \tag{5.82}$$

According to the Chapman-Enskog analysis, the collision frequency ω is related to the dynamic viscosity [95]. In that sense the relaxation time $\tau = 1/\omega$ is given as

$$\tau = 3\,\nu_{(\mathrm{lb})} + \frac{1}{2} , \tag{5.83}$$

where the factor ½ is added with respect to a finite time step width. Finally, the scaling for the density ρ_0 can be freely chosen, whereas the physical pressure is linked via the equation of state as

$$p_{(\mathrm{p})} = \frac{1}{3} p_{(\mathrm{lb})} \frac{\delta_x^2}{\delta_t^2} \rho_0 . \tag{5.84}$$

As already outlined above in sec. 5.3.2 the viscosity in eq. (5.21) must not be negative, thus as the lower limit, the relaxation time should be $\tau > 0.5$. To recap on the choice of the spatial resolution N_x and the reference velocity in lattice units $u_{0,(\text{lb})}$ it gets obvious, that both values have an impact on the stability in terms of the relaxation time τ.

Furthermore, since the lattice Boltzmann method does not support high Mach number flows[7], the magnitude of the lattice velocity is limited by the speed of sound c_s. However, the lattice Boltzmann method solves for quasi compressible flows, where the numerical accuracy is affected by compressibility effects. According to the comprehensive derivation in [121], while the compressibility effects scale with $\text{Ma}^2 \sim \delta_t^2/\delta_x^2$, the lattice error scales with δ_x^2 since the lattice Boltzmann method is second order accurate. This leads to a trade-off between lattice error and compressibility error, and gives the common relation $\delta_t \sim \delta_x^2$, cf. [72, 121, 215].

For illustration, sample parametrisation is carried out in appendix E, and the corresponding input parameters for the simulation with PACE3D are given.

5.10 Validation

For validation of the above outlined lattice Boltzmann method integrated in PACE3D three different testcases are performed throughout the subsequent sections, where simulations are conducted for sharp as well as diffuse interface setups and the results are compared to analytic or reference data from literature.

5.10.1 Plain Poiseuille-flow

The velocity profile of the plain Poiseuille-flow is given as

$$u(y) = \frac{dp}{dx} \frac{1}{2\mu} \left(\left(\frac{h}{2} \right)^2 - y^2 \right) \tag{5.85}$$

where the origin of the coordinate system is placed at the centerline of the channel, h is the channel height and μ the dynamic viscosity. In

---

[7]Note, there are flavours of models and methods capable of high Mach number or even supersonic flows [30, 198, 215], which are not within the scope of this work.

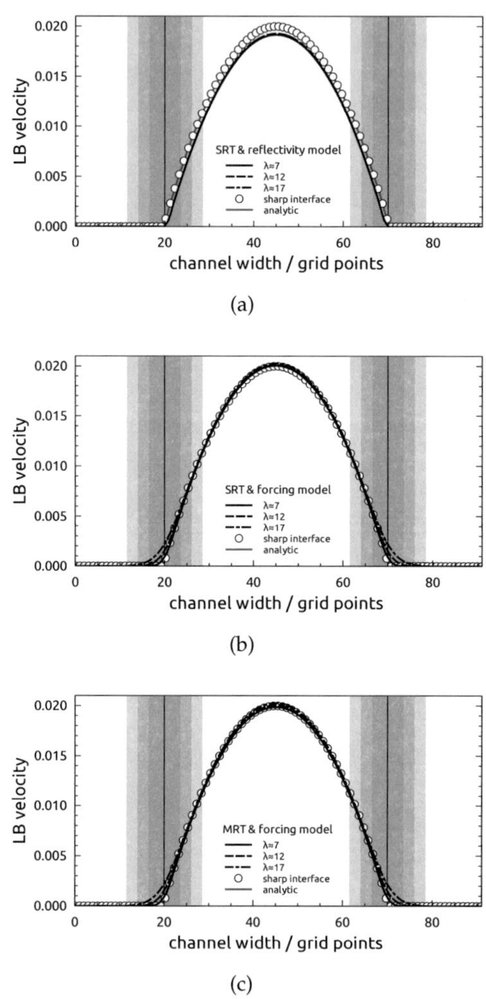

Figure 5.3: Comparison of velocity profiles of the plain Poiseuille flow obtained analytically, with sharp interface with different diffuse interface and lattice models: (a) SRT lattice BGK model and *reflectivity* model (b) SRT lattice BGK model and *forcing* model (c) MRT lattice Boltzmann model and *forcing* model. The shadowy regions mark the diffuse interface layer.

general, the flow is driven by a pressure gradient dp/dx in flow direction. For a gravity driven flow, we replace $dp/dx = \rho g$.

Figure 5.3 depicts velocity profiles obtained for different lattice models and diffuse interface couplings in comparison with the corresponding analytical and the sharp interface solutions. The channel width is discretised using 50 lattices starting at grid coordinate 20 and ending at 70 for the sharp interface setup. For all diffuse interface setups, a smooth distribution of the order parameter is chosen, such that $\phi = 0.5$ is located at the grid coordinate 20 and 70, respectively. Different diffuse interface widths are tested, where the dimensionless control parameter $\varepsilon = 3$, 5 and 7 corresponds to the interface widths $\lambda \approx 7$, 12 and 17, where $\lambda = \varepsilon\,\pi^2/4$ in grid units.

While the sharp interface solution matches the analytical profile, we observe differences in the velocity distribution for the *reflectivity* model, cf. fig. 5.3(a). The magnitude of the velocity profile is at maximum 4% lower than the reference solution. However, the results of the *forcing* model are in very good agreement with the analytic solution for both, the SRT as well as MRT lattice Boltzmann method respectively, cf. figs. 5.3(b) and 5.3(c).

5.10.2 Lid driven cavity

For more than three decades [24], the lid-driven cavity flow is a well known and most probably one of the most implemented benchmark problems in computational fluid dynamics, cf. [61, 62, 84]. Despite its simplicity, which makes the problem appropriate for coding and numerics, it retains plenty of flow physics, such as singularities at two of its corners and counter rotating vortices.

The cavity is spatially discretised according to the stability requirements [216] and with respect to the different Reynolds numbers performed with 100 × 100 cells for Reynolds numbers Re=100 and Re=1000. For a Reynolds number of Re=5000 the sharp interface solution using the MRT model required a resolution of 250 × 250 cells, whereas the stability is obviously positively influenced by employing the MRT forcing model where 150 × 150 cells lead to a accurate and stable solution, cf. figs. 5.4. As a consequence of the plain Poiseuille test, we exclusively apply the *forcing* model to this testcase. The normalised velocity profiles along the vertical and the horizontal center line are displayed in figs. 5.4

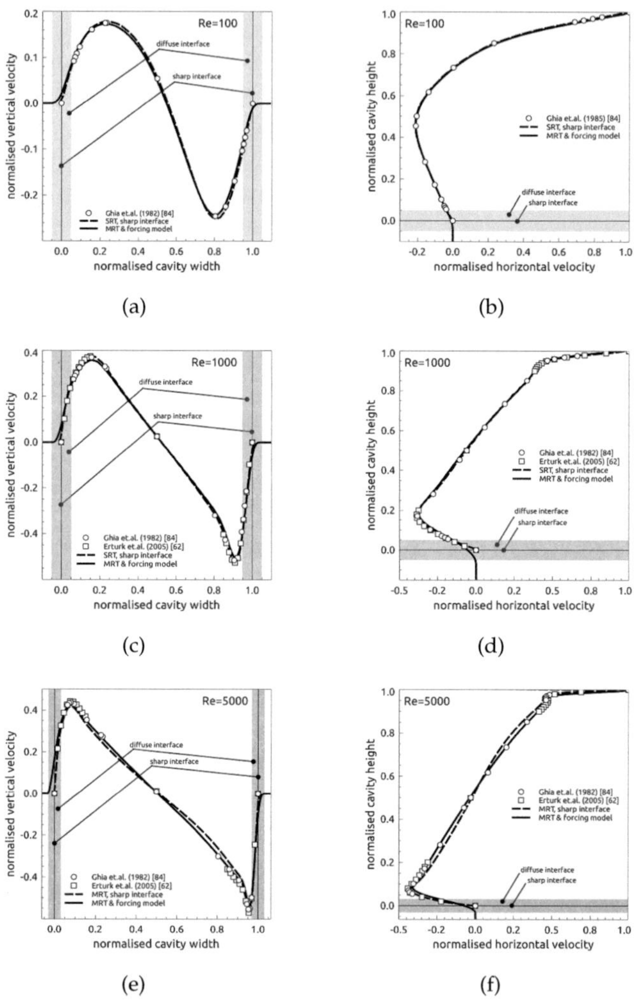

Figure 5.4: Velocity profiles for the *lid driven cavity* testcase: normalised vertical velocity along horizontal centerline for Reynolds numbers (a) Re=100, (c) Re=1000 and (e) Re=5000, and normalised horizontal velocity along vertical centerline for Reynolds numbers (b) Re=100, (d) Re=1000 and (f) Re=5000. The shadowy regions mark the diffuse interface areas.

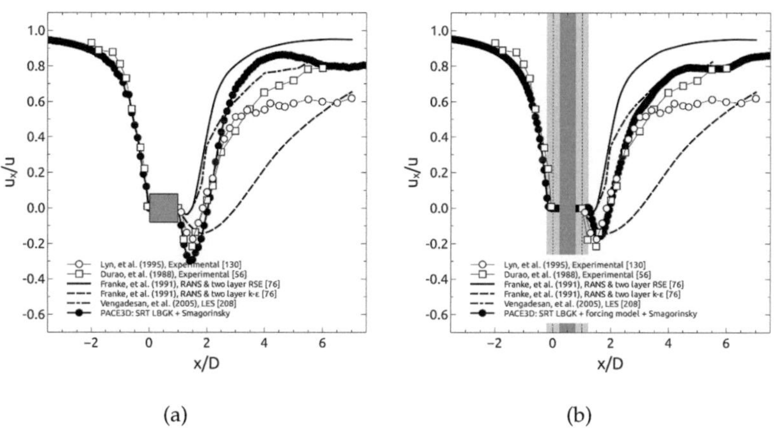

(a) (b)

Figure 5.5: Time averaged normalised streamwise velocity along horizontal centerline for the turbulent flow past square cylinder.

for Reynolds numbers Re=100, 1000 and 5000, along with the reference solutions from [62, 84].

For the Reynolds numbers Re=100 and Re=1000 we observe minor deviations in the velocity profiles in horizontal centerline direction between sharp and diffuse interface solutions, cf. figs. 5.4(a) and 5.4(c), whereas the differences for the velocity distribution along vertical centerline direction are negligible, cf. figs. 5.4(b) and 5.4(d). For a high Reynolds number Re=5000, the solution is subject to small oscillations, indicating transient effects which cause more obvious but still acceptable differences, as depicted in figs. 5.4(e) and 5.4(f).

5.10.3 Square cylinder in high Reynolds number flow

The simulation of flow past square cylinder [71, 124, 190] is performed for flow at Re=22'000 using the afore mentioned SRT lattice Boltzmann BGK method employing the Smagorinsky turbulence model. A square cylinder of dimensions $D \times D$ is located at the center of a computational domain of extent $-10\,D \leq x \leq 20\,D$ in streamwise direction and $-10\,D \leq y \leq 10\,D$ in lateral direction. The velocity boundary condition is applied at the

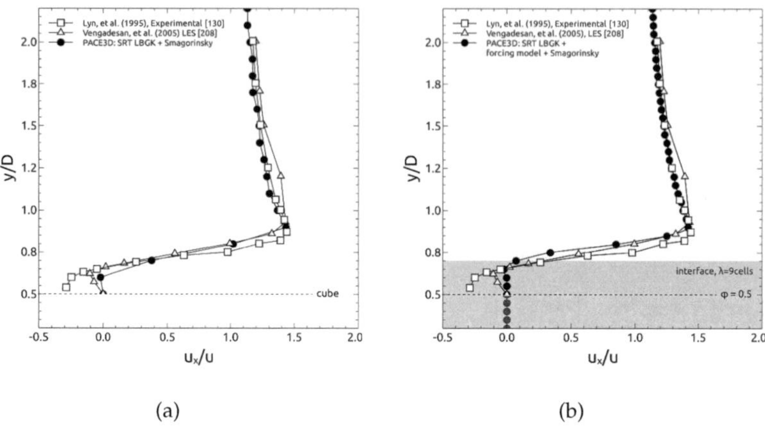

(a)   (b)

Figure 5.6: Time averaged normalised streamwise velocity along vertical center-line for the turbulent flow past square cylinder.

inlet, pressure boundary condition at the outlet, and no slip boundary conditions on the outer walls. Providing that the flow is two-dimensional, periodic boundary conditions are composed in lateral direction. The Reynolds number is defined as

$$\mathrm{Re} = \frac{u\,D}{\nu} \tag{5.86}$$

using the uniform inlet velocity U, the width of the square cylinder D and the kinematic viscosity ν. A spatial discretisation of 200×300 cells is chosen for the sharp interface setup, whereas a finer resolution of 400×600 cells is used for the diffuse interface setup, in order to reasonably resolve the diffuse boundaries.

In fig. 5.5 the time averaged streamwise velocity along the wake centerline is compared to experimental and numerical reference data, cf. [56, 76, 124, 130, 208]. The results of our method are in the same range and at the same level as the experimental and numerical reference data. While the sharp interface solution (–•–) in fig. 5.5(a) slightly overpredicts the negative peak in the wake of the cube compared to the experimental data (–○– and –□–), the diffuse interface results are in very good agreement, cf.

fig. 5.5(b). The velocity slope in the wake is well mapped by both, the sharp as well as the diffuse interface solution.

The velocity profile in lateral direction at $x/D = 0.5$ is given in figs. 5.6 for the sharp and diffuse interface solutions in comparison to experimental data. For both, the bulk velocity profile is in very good agreement with the experimental data, and also the slope is well mapped. Due to the prerequisite of a uniform Cartesian grid for the lattice Boltzmann simulation, and a limited reasonable grid resolution, the details of the flow in the immediate vicinity of the surface lacks from grid resolution for the sharp interface setup, cf. fig. 5.6(a). For a finer mesh using the diffuse interface approach, the details are smoothed intrinsically by the dissipative forcing inside the interface layer, cf. fig. 5.6(b).

Chapter 6

Diffuse interface heat transfer[1]

With respect to heat transfer across a phase boundary, it has been shown that for different thermal conductivities of two phases which are in contact, a scalar interpolation of this quantity causes the phenomena of *heat trapping* across a non-stationary interface, which is caused by unequal fluxes of heat current across two sides of an interface [3, 140]. The defect manifests itself as a discontinuity of the temperature when asymptotes of the bulk temperature fields are extended on to the interface from both phases. A similar phenomenon has also been found in mass transfer across the interface between phases which have differing diffusivities, the result in this case being a *solute trapping* causing a chemical potential jump at the interface [111]. In the limit of vanishing diffusive current through the bulk solid, it can be shown through a formal asymptotic analysis this jump is proportional to the velocity of the interface. A solution was proposed through the formulation of an anti-trapping current [111], which corrects for this defect though a current from the solid (zero diffusivity) to the liquid that is proportional to the velocity.

In the limit of non-vanishing diffusivity in the solid, the flux of heat/-mass across the interface can be present even for non-zero velocities. Due to the non-homogeneity of the thermal conductivity across the interface an incorrect interpolation leads to a chemical potential/temperature at the interface, which is *independent* of the velocity of the interface [93, 169]. [163] correct for this defect, when the interface is non-stationary, by modifying the anti-trapping current by a factor which depends on the flux reaching the solid side of the interface. The problem is however unsolved for stationary interfaces.

Physically, the problem can be seen as the *retrieval* of the correct effective thermal *resistance* for the different currents across the phase-field interface. Essentially, for currents flowing in a direction normal to

---

[1]Parts of the subsequent sections are submitted for publication in *Modelling and Simulation in Materials Science and Engineering* [68, 70].

the interface, the two phases are arranged in a serial manner, rendering the effective resistance as $R = R_\alpha \phi_\alpha + R_\beta \phi_\alpha$, where the ϕ_α and ϕ_β are the indicator functions representing the volumes of the two phases at a given point. Clearly, since the current is given by $k\nabla T$, k being the thermal conductivity, the resistance is given by $1/k$, and therefore, the correct interpolation of the thermal conductivity would read $1/k_\perp = \phi_\alpha/k_\alpha + \phi_\beta/k_\beta$. However, for currents flowing in the plane of the interface parallel to it, the phases are arranged parallely, thereby the correct interpolation should be $k_\parallel = k_\alpha \phi_\alpha + k_\beta \phi_\beta$. This duality cannot be treated in a scalar framework and [158] formulates a tensorial approach where the thermal conductivity can be written as a matrix such that,

$$\begin{pmatrix} j_n \\ j_t \end{pmatrix} = \begin{pmatrix} k_\perp & 0 \\ 0 & k_\parallel \end{pmatrix} \begin{pmatrix} \nabla T_n \\ \nabla T_t, \end{pmatrix} \tag{6.1}$$

where j_n and j_t are the currents normal and parallel to the interface.

While the above construction solves the asymmetry in fluxes at steady-state, the transient evolution of the temperature field across an in-homogeneous interface has not been treated. The relevant parameter related to the change of temperature for a unit change in the internal energy is indeed related to the heat capacity of the interface. In this chapter we outline the problems arising with the different interpolation schemes for the heat capacity, and thereby derive a modified divergence operator constructed out of the effective capacities of the interface, for the different currents across the interface. In this endeavour, we draw upon inferences from the analogical variable in electrical circuits which is the electrical capacitance.

Recent developments [83] indicate, that there are more sophisticated methods for the solution of the model problems we are dealing with in the subsequent section. We are aware of the fact, that non moving interfaces without phase transition or morphological processes, are accurately and in particular more efficiently covered by means of classical finite difference or finite element methods. The elegance of the phase-field method, lies in being able to treat problems of complex geometries which would otherwise require very efficient numerical schemes in the FEM, FVM framework, through relatively less intensive computational methods, thus allowing for ease of parallelisation required for simulating larger domains.

In the course of fluid flow and heat transfer applications we utilise the phase-field method for the modelling of transient heat conduction for composite materials, thus paying attention on the correct representation of the underlying physical processes. In the following, the method proposed in [158] is extended to the general three dimensional and transient case in the context of heat conduction.

6.1 Physical Model

We will start from the well known heat conduction equation [167]

$$\rho\, c_V\, \partial_t T = \langle \nabla , q \rangle,\tag{6.2}$$

with time derivative $\partial_t T$, density ρ, specific heat c_V, and heat flux vector q. The latter is proportional to the product of the thermal conductivity k and the gradient of the temperature potential

$$q = k\, \nabla T.\tag{6.3}$$

We consider a computational domain Ω with the continuous bulk phases α and β, which are separated by a interface of finite width, cf. fig. 4.1. An equally spaced uniform Cartesian grid, which is not aligned with the interface is employed. In the bulk, the material properties k and c_V are constant, i.e. k^α, k^β and c_V^α, c_V^β according to the phases α and β respectively. Inside the transition region of the diffuse interface layer, the material properties are supposed to be interpolation functions of the order parameters, such that they maintain the physics in the bulk and asymptotically recover the physics of a sharp interface solution. Thus we define $k(\phi)$ as the interpolation function for the thermal conductivity and $C_V(\phi)$ as the interpolation function for the volumetric heat capacity $C_V(\phi) = \rho\, c_V(\phi)$.

Thus, the basic scalar temperature evolution equation we consider is

$$C_V(\phi)\, \partial_t T = \langle \nabla, | k(\phi)\nabla T \rangle.\tag{6.4}$$

By default, the initialisation of the phase distribution is done by a sharp interface setup. At first we performed a phase-field simulation without driving forces, in order to establish a diffuse interface with finite width λ. The pre-calculations were stopped as the phase-field landscape became stationary.

In the following, the functions $k(\phi)$ and $C_V(\phi)$ need to be defined. As outlined in ch. 4, the material properties of the bulk phases are interpolated across the diffuse interface. The derivation of an eligible interpolation formalism is essential, in order to perform quantitative simulations. Against this background, the interpolation of mobility and capacity coefficients of heat conduction by means of a general diffusive transport process are developed in the following sections. In avoidance of unnecessarily convoluted examples, we focus on two phase systems with non-moving steady interfaces. Thus, the stationary phase field ϕ_α is solely used to distinguish between the different phases.

6.2 Classical sharp-interface methodology

Already [168] is concerned with the modelling of the mobility coefficient of a diffusive process at the interface of a composite material in the context of a sharp-interface finite volume (FV) model. It is derived, that the mobility coefficients are represented best using a harmonic interpolation. We give a brief outline of the concept.

Consider the control volume approach depicted in fig. 6.1, where the conductivities k_L and k_R are defined at cell centers, with respect to the phases 'L' and 'R' respectively. For discretisation purposes we need to

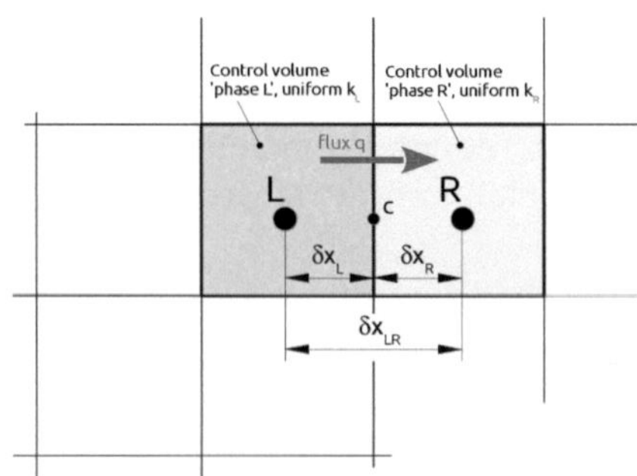

Figure 6.1: Sharp interface finite volume example.

express the conductivity at interface position 'c'. A first attempt might be to use a linear variation of $k(x)$ between the values of k_L and k_R respectively, which leads to

$$k_c = \frac{\delta x_R}{\delta x_{LR}} k_L + \frac{\delta x_L}{\delta x_{LR}} k_R . \tag{6.5}$$

If the interface 'c' is located midway between 'L' and 'R', $\delta x_R / \delta x_{LR} \rightarrow 0.5$, and k_c is the arithmetic mean of k_L and k_R. As we can see from the following survey, this is not valid, neither accurate nor can it handle abrupt changes in k.

The conductivity k is not important, since our main objective is a good representation of the heat flux q_c at the interface. It is desired, that interpolation of k leads to correct q_c. According to Fick's law the heat flux q is proportional to a temperature gradient $\Delta T / \Delta x$, with the proportionality factor k, the thermal conductivity. Assuming that the conductivities are constant throughout the cells, and with the temperature difference with respect to the distance to the interface c, the heat fluxes are defined for both sides of the interface

$$q_L = \frac{k_L}{\delta x_L}(T_L - T_c) \quad \text{and} \quad q_R = \frac{k_R}{\delta x_R}(T_c - T_R), \tag{6.6}$$

where T_c is the interface temperature, and rearranged

$$T_c = T_L - q_L \frac{\delta x_L}{k_L} = q_R \frac{\delta x_R}{k_R} + T_R . \tag{6.7}$$

From the continuity condition $q_L = q_R = q_c$, we get

$$T_L - q_c \frac{\delta x_L}{k_L} = q_c \frac{\delta x_R}{k_R} + T_R \tag{6.8}$$

and obtain

$$q_c = \frac{T_L - T_R}{\left(\frac{\delta x_R}{k_R} + \frac{\delta x_L}{k_L} \right)} \tag{6.9}$$

Now we model the conduction equation for point 'c' as

$$k_c = \frac{q_c \, \delta x_{LR}}{T_L - T_R} = \frac{T_L - T_R}{T_L - T_R} \frac{\delta x_{LR}}{\left(\frac{\delta x_R}{k_R} + \frac{\delta x_L}{k_L} \right)} = \frac{\delta x_{LR}}{\left(\frac{\delta x_R}{k_R} + \frac{\delta x_L}{k_L} \right)} \tag{6.10}$$

and write

$$\frac{1}{k_c} = \frac{\delta x_L}{\delta x_{LR}}\frac{1}{k_L} + \frac{\delta x_R}{\delta x_{LR}}\frac{1}{k_R} \tag{6.11}$$

So, k_c is the harmonic mean, and the ratios $\delta x_R/\delta x_{LR}$ and $\delta x_R/\delta x_{LR}$ are the weighting functions, depending on the position of the interface.

We can further test two limits, where we assume that the interface is right in the middle of the left and the right cell. If one conductivity tends to zero, e.g. $k_R \to 0$, this leads to

$$\lim_{k_R \to 0} k_c = 0 \tag{6.12}$$

representing an insulator. If we assume $k_L \gg k_R$, we get

$$\lim_{k_L \to \infty} k_c = \frac{k_R\,\delta x_{LR}}{\delta x_R} \tag{6.13}$$

Here, the interface conductivity does not depend on k_L, since there is no resistance from phase 'L'. However, $k_c \neq k_R$ since q_c is to be captured, and thus $k_c = k_R\,\delta x_{LR}/\delta x_R$ to account for the nominal distance.

6.3 Capacitance analogy

The utilisation of the diffuse-interface methods for problems of transport in non-homogeneous media, requires careful understanding of the transport mechanisms across the interface which are intrinsically related to the interpolation of the related material properties across the interface. In heat transfer problems the material parameter related to the flux of heat across the interface is the thermal conductivity. In [158] the authors describe a tensorial approach for removing the temperature jump discontinuity at an interface, which is seen in models using a scalar interpolation of the thermal conductivity that are unable to capture the effective surface resistance in multi-dimensional flow of current. While, the approach allows for the correct boundary conditions of the temperature field at steady state, the case of transient heat flow across an interface is untreated.

In the subsequent sections we intend to contribute an approach for the time-dependent problem, the relevant material parameter here being the heat capacity. To arrive at the correct interpolation scheme for this material property, we draw on an analogy to the effective capacitance in electric circuits. For this, consider the similarity of the internal energy e and electric charge Q, where infinitesimal changes in either quantity can be related to the corresponding change in temperature T and voltage difference V respectively as,

$$\delta e = C_V \, \delta T$$
$$\delta Q = C_e \, \delta V \tag{6.14}$$

where the temperature potential T corresponds to the electric potential V as well as the heat capacity C_V corresponds to the electric capacitance C_e.

It is well known, that the effective capacitance of an electrical circuit depends on the direction of the current with respect to a serial or parallel arrangement of the capacitors. Therefore, let us consider the effective capacitance in terms of the capacitors formed of material α and β. A serial arrangement corresponds to a current normal to an interface between the capacitors, for which the effective capacitance writes as $1/C_{e,\perp} = 1/C_e^\alpha + 1/C_e^\beta$. On the other hand, for a parallel arrangement, the potential across both phases is the same, and the effective capacitance writes as $C_{e,\parallel} = C_e^\alpha + C_e^\beta$.

In the following sections, we first show the discrepancies that arise with the different interpolation schemes and thereafter employ this analogy and derive the interpolation functions for the heat capacity of the diffuse interface approach with respect to the direction of the heat currents according to the interface.

6.4 Numerical survey on 1D test case

In this section we present successive numerical investigations on the transient heat conduction problem using the concept of the diffuse interface of the phase-field method [154], applying the methodology of [158] for the thermal conductivity $k(\phi)$ and applying different interpolation schemes for the volumetric heat capacity $C_V(\phi)$ with respect to eq. (6.4). For the sake of simplicity and computational effort, a rather simple setup of a quasi one-dimensional composite material is used to investigate the diffuse interface transient heat conduction problem, cf. fig. 6.2. The composites are equally spaced in a physical domain of length $0.05\,\mathrm{m}$ which is resolved by 200 cells in z direction, where the temperature is held constant at the left boundary at $z = 0$ while the right boundary at $z = z_{max}$ is set to adiabatic. In the sense of a quasi one-dimensional setup, the other directions are resolved by just one cell and the corresponding boundaries are subject to adiabatic conditions. The stationary interface is located at the center of the domain.

Starting from a constant initial temperature of $293\,\mathrm{K}$ in the interior of the domain, we instantaneously apply a temperature of $303\,\mathrm{K}$ at the left boundary at $z = 0$. Since we are interested in high ratios of thermal conductivity and heat capacity, aluminium and air are used as materials. At this point we have to indicate, that natural convection is not considered. Only diffuse thermal transport is modelled. The physical properties used for the simulations are listed in tab. 6.2.

Qualitative and quantitative comparison in the sense of model development is achieved with simulations performed applying the commercial

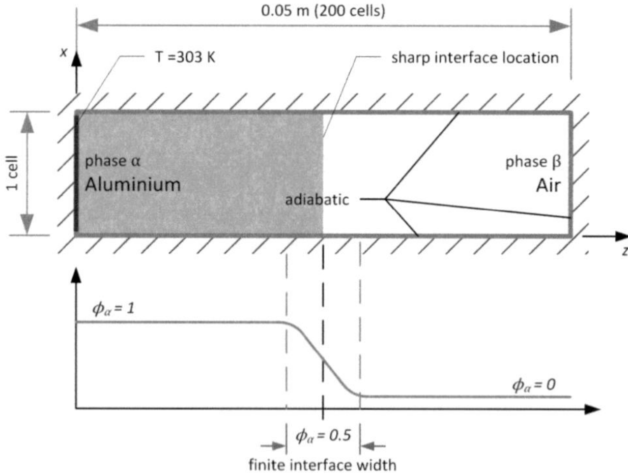

Figure 6.2: Outline of the one dimensional test case together with an indication for the profile of the order parameter ϕ_α. The sharp interface location corresponds to $\phi_\alpha = \phi_\beta = 0.5$.

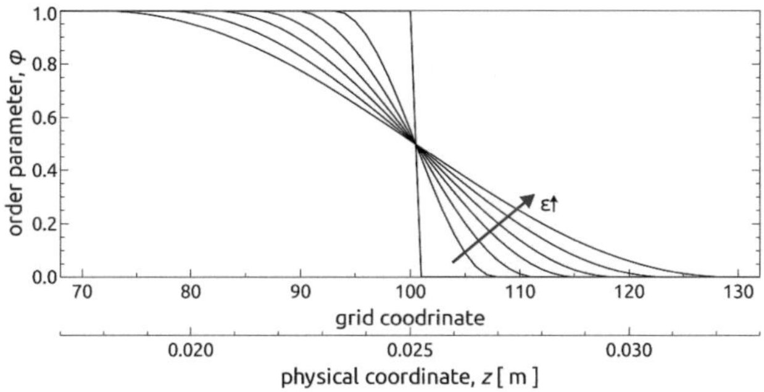

Figure 6.3: Phase-field profile shape for different interface widths.

package StarCCM of CD-Adapco [27], and setting an identical numerical domain and configuration. Validation of the final chosen model against analytical solutions are provided in section 6.6.

Unless stated otherwise, we use an interface of about 15 cells. Based on the steady state solution for the order parameter [154], the interface width $\lambda = \epsilon \pi^2/4$, where ϵ is a model parameter of the phase field method [154]. Here, the location of the sharp interface is represented by $\phi_\alpha = \phi_\beta = 0.5$, cf. fig. 6.2. For studying the influence of interface width, the parameter ϵ is increased up to 21, which is equivalent to about 53 cells. The different interfaces are depicted in fig. 6.3. The profile $\epsilon = 0$ is used to generate sharp interface like solutions for comparison and validation.

Starting with the 1D representation, both, the sharp interface as well as the diffuse interface models refer to the interfacial conditions of matching temperatures approaching from both sides of the interface, given by

$$T^\alpha = T^\beta. \tag{6.15}$$

Furthermore, the setup corresponds to the condition reading

$$k^\alpha \nabla_\perp T = k^\beta \nabla_\perp T = q_\perp = \text{const}. \tag{6.16}$$

In the subsequent sections we perform simulations on heat conduction perpendicular to the interface. The quasi one-dimensional setup allows us to employ the heat conduction eq. (6.4) with scalar mobility and capacity coefficients. While we use the harmonic interpolation for the thermal conductivity $k(\boldsymbol{\phi})$, we will study the effect of applying different interpolation schemes for the specific heat $C_V(\boldsymbol{\phi})$, in order to correctly render the temperature evolution in the bulk phases.

At this point, it is important to note, that in the context of the phase field model, we intentionally choose to model an nonphysical finite interface at a mesoscopic scale. Doing so, the method avoids the explicit definition of a coupling (boundary) condition at the interface. We are aware that we will receive deviations inside the interface, compared to sharp interface solution, and we refrain from a physical interpretation since the concept of a finite interface is a mathematical model.

6.4.1 Sharp interface solutions

At first we confirm, that the modelling and discretisation of the governing bulk conservation equations of heat conduction give reasonable results. Therefore, we perform calculations using a unit step function instead of a diffuse interface representation. Doing this, the order parameter ϕ changes discontinuously from 0 to 1 from one cell to the next.

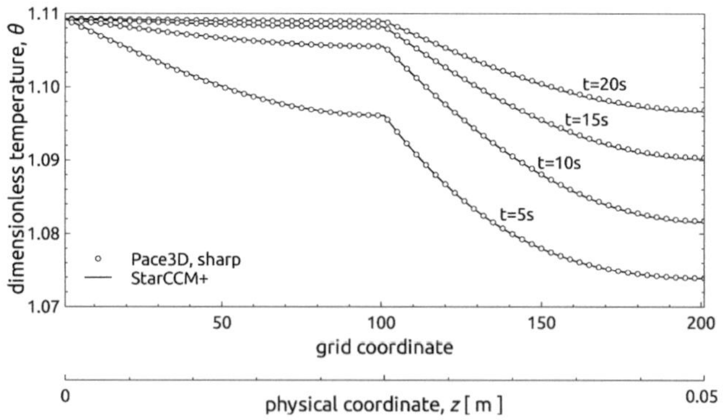

Figure 6.4: Distributions of dimensionless temperature at different times, calculated with StarCCM (sharp interface) and with PACE3D using a unit step function in ϕ_α.

Due to the discretisation, the temperature gradient of the heat flux $k\nabla T$ is evaluated at the cell faces. Thus, the thermal conductivities k need to be interpolated. In case of the unit step function, evaluation of the inverse interpolation is written as

$$\frac{1}{k(\boldsymbol{\phi})} = \sum_{\alpha} \frac{h(\phi_{\alpha})}{k_{\alpha}} \tag{6.17}$$

at the cell face implying $\phi_{\alpha} = 0.5$. Herein, and in the following, the function $h(\phi)$ represents an interpolation function, which is used to smooth or sharpen the distribution of ϕ inside the interface, whereas the bulk values are recovered. In the subsequent sections h^{α} and h^{β} are used as an abbreviation for $h(\phi_{\alpha})$ and $h(\phi_{\beta})$, respectively. Thus the interpolation recovers the harmonic mean of the thermal conductivities, cf. sec. 6.2.

Using the unit step function, we evaluate the 1D testcase at four different instants of time: 5, 10, 15 and 20 s. Figure 6.4 depicts the distribution of normalised temperature along the z-coordinate, where the simulations done with PACE3D perfectly renders the temperature distributions at different times compared to StarCCM.

6.4.2 Linear interpolation in volumetric heat capacity

By default we start with a linear (direct or arithmetic) interpolation of the volumetric heat capacity inside the diffuse interface region

$$C_V(\boldsymbol{\phi}) = \sum_{\alpha} C_V^{\alpha} h(\phi_{\alpha}). \tag{6.18}$$

The combination of inverse interpolation of $k(\boldsymbol{\phi})$ and direct interpolation of $C_V(\boldsymbol{\phi})$ respectively, leads to a kind of 'barrier' in the distribution of the locally derived thermal diffusivity $a(\boldsymbol{\phi}) = k(\boldsymbol{\phi})/C_V(\boldsymbol{\phi})$, cf. fig. 6.5(a). This causes a reduced diffusive transport, and a steep temperature gradient inside the diffuse interface, cf. fig. 6.5(b). Thus, the direct interpolation in volumetric heat capacity is not suitable for the problem of heat flux perpendicular to the interface.

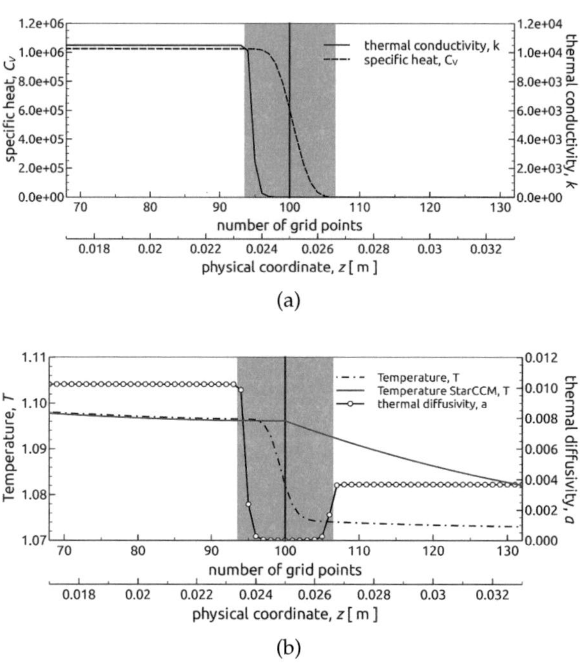

Figure 6.5: Inverse/harmonic interpolation in k and direct/arithmetic interpolation in C_V. (a) interface profile ϕ together with material properties k and C_V, (b) temperature profile, interface profile ϕ and thermal diffusivity a. The shadowy area indicates the diffuse interface region, with a vertical line at $\phi = 0.5$.

6.4.3 Inverse interpolation in volumetric heat capacity

Consequently, we apply the inverse/harmonic interpolation for both, $k(\boldsymbol{\phi})$ and $C_V(\boldsymbol{\phi})$, i.e.

$$\frac{1}{C_V(\boldsymbol{\phi})} = \sum_\alpha \frac{h(\phi_\alpha)}{C_V^\alpha} \,. \tag{6.19}$$

Referring to fig. 6.6(a), the distributions of k and C_V qualitatively coincide. This results in a quite promising temperature distribution that gives a qualitatively good picture but nevertheless does not coincide with the reference curve obtained with StarCCM, cf. fig. 6.6(b).

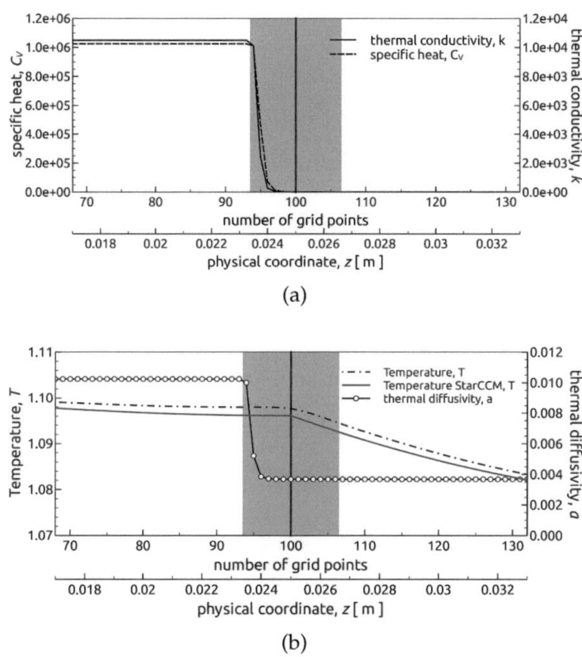

Figure 6.6: Inverse/harmonic interpolation in k and in C_V. (a) interface profile ϕ together with material properties k and C_V, (b) temperature profile, interface profile ϕ and thermal diffusivity a. The shadowy area indicates the diffuse interface region, with a vertical line at $\phi = 0.5$.

Since the evaluated temperature is higher than that of the reference curve, the temporal evolution of temperature apparently is too fast. The same is noticed for different instances of time, which are not depicted here. This gets even more reasonable when trying to physically explain the effect of the diffuse interface by smoothing the spatial transition from one phase to the other. Approaching the interface from e.g. the bulk of phase α we are still in phase α with decreased conductivity and heat capacity of the concerned phase. Since $\partial_t T \propto 1/C_V$, the actual temporal evolution is faster. At the same time, also the flux $q \propto k$ is decreased, which seems to balance the problem. Obviously this is not sufficient to capture the physically correct temporal temperature evolution in detail.

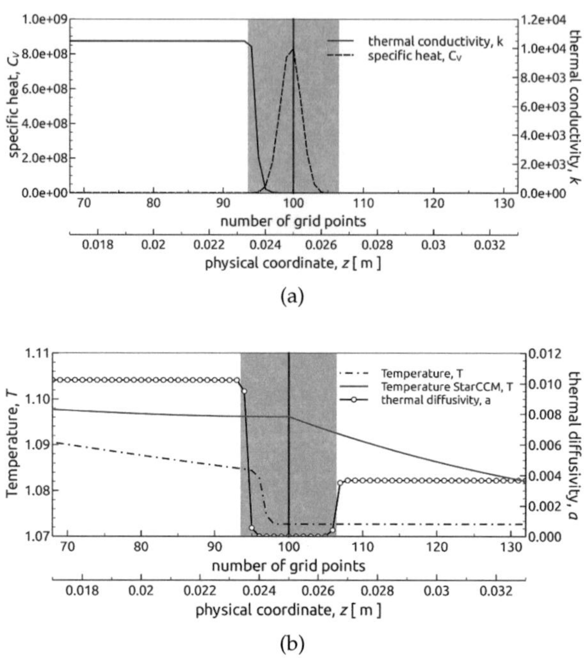

Figure 6.7: Inverse/harmonic interpolation in k and evaluation of C_V by linear/direct interpolation of thermal inertia I. (a) interface profile ϕ together with material properties k and C_V, (b) temperature profile, interface profile ϕ and thermal diffusivity a. The shadowy area indicates the diffuse interface region, with a vertical line at $\phi = 0.5$.

6.4.4 Direct interpolation in thermal inertia

Referring to the semi-infinite contact problem [9], we will make use of the so called thermal inertia I, which is a composed material property, characterising the importance of dynamic effects. A higher thermal inertia will cause a system to take longer to reach thermal equilibrium. It is defined as square root of the product of thermal conductivity and volumetric heat capacity in the form

$$I = \sqrt{k\,C_V} \tag{6.20}$$

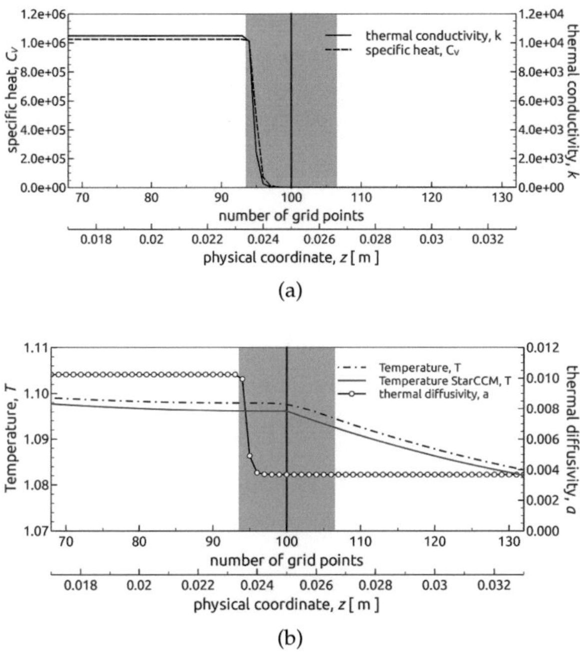

Figure 6.8: Inverse/harmonic interpolation in k and evaluation of C_V by inverse/harmonic interpolation of thermal inertia I. (a) interface profile ϕ together with material properties k and C_V, (b) temperature profile, interface profile ϕ and thermal diffusivity a. The shadowy area indicates the diffuse interface region, with a vertical line at $\phi = 0.5$.

Here we derive the interpolation of C_V via the direct interpolation of the thermal inertia

$$I(\phi) = \sqrt{k(\phi)\, C_V(\phi)} = I^\alpha h^\alpha + I^\beta h^\beta. \tag{6.21}$$

We substitute $k(\phi)$ by the inverse interpolation and define

$$C_V(\phi) = \frac{\left(I^\alpha h^\alpha + I^\beta h^\beta\right)^2}{k(\phi)} \tag{6.22}$$

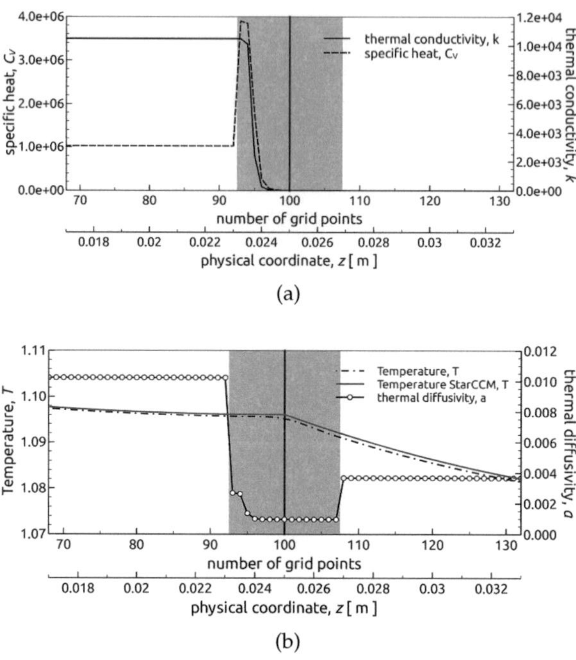

Figure 6.9: Inverse/harmonic interpolation in k and factorised inverse/harmonic interpolation of C_V using a single fixed factor. (a) interface profile ϕ together with material properties k and C_V, (b) temperature profile, interface profile ϕ and thermal diffusivity a. The shadowy area indicates the diffuse interface region, with a vertical line at $\phi = 0.5$.

The contour lines of the functions $k(\phi)$ and $C_V(\phi)$ across the interface are plotted in fig. 6.7(a). Again, as for the direct interpolation of C_V in section 6.4.2, we get a kind of barrier, i.e. due to the interpolation routine, the thermal diffusivity reaches very small values inside the interface, which inhibits the evolution of temperature profile, as depicted in fig. 6.7(b).

6.4.5 Inverse interpolation in thermal inertia

Consequently, we do the inverse/harmonic interpolation for the thermal inertia I

$$\frac{1}{I(\boldsymbol{\phi})} = \frac{1}{\sqrt{k(\boldsymbol{\phi})C_V(\boldsymbol{\phi})}} = \frac{h^\alpha}{I^\alpha} + \frac{h^\beta}{I^\beta}. \tag{6.23}$$

and combine this expression with the inverse interpolation of $k(\boldsymbol{\phi})$ reading

$$C_V(\boldsymbol{\phi}) = \frac{1}{k(\boldsymbol{\phi})} \left(\frac{h^\alpha}{I^\alpha} + \frac{h^\beta}{I^\beta} \right)^{-2}. \tag{6.24}$$

As depicted in fig. 6.8 we get comparable results as using the inverse/harmonic interpolation for k and C_V. Even though both, the direct and inverse interpolations in thermal inertia do not seem to deviate, we included the results for thoroughness and comprehension.

6.4.6 Factorised inverse interpolation in volumetric heat capacity

In the preceding test cases we have seen that we derive the best agreement with the sharp interface solution for the case when the heat capacity is inverse interpolated. A motivation for this has already been discussed before in sec. 6.3, in that the effective capacitance is derived through an inverse interpolation of the heat capacity for the case where the current is normal to the interfacial direction. However, in doing so, we cannot re-derive the volume integral internal energy across the interface. In the following, we present a factorised scheme where we artificially modify the heat capacity in order to derive the correct volume integral of the internal energy.

We start from the formulation of an effective heat capacity for the sharp interface, $\hat{C}_V$, in a domain which is equally occupied by phases α and β. Using the volume fractions χ_α and χ_β we can write

$$\hat{C}_V = \chi_\alpha C_V^\alpha + \chi_\beta C_V^\beta. \tag{6.25}$$

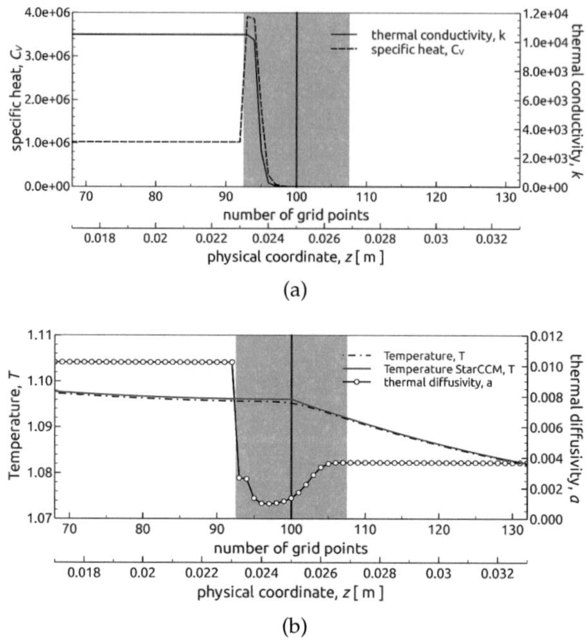

Figure 6.10: Inverse/harmonic interpolation in k and factorised inverse/harmonic interpolation of C_V using a single decaying factor. (a) interface profile ϕ together with material properties k and C_V, (b) temperature profile, interface profile ϕ and thermal diffusivity a. The shadowy area indicates the diffuse interface region, with a vertical line at $\phi = 0.5$.

Since the domain is separated in exactly two parts, we consider $\chi_\alpha = \chi_\beta = \frac{1}{2}$, and obtain the arithmetic mean

$$\hat{C}_V = \frac{1}{2}\left(C_V^\alpha + C_V^\beta \right). \tag{6.26}$$

To derive an effective volumetric heat capacity for the diffuse interface, $\tilde{C}_V$, we define an integral heat capacity distribution $C_V'(\phi)$ over the not yet known interface in the range of the diffuse interface width by

$$\tilde{C}_V = \frac{1}{\lambda} \int_{-\lambda/2}^{\lambda/2} C_V'(\phi)\, dx. \tag{6.27}$$

if we assume, that the effective heat capacities are equal for the sharp and diffuse interface profile, we can set $\check{C}_V = \tilde{C}_V$ and obtain

$$\frac{\lambda}{2}\left[C_\alpha + C_\beta\right] = \int_{-\lambda/2}^{\lambda/2} C_V'(\phi)\,dx. \tag{6.28}$$

As discussed above, the inverse harmonic interpolation (HM) for C_V shows promising results to establish the expected temperature profiles across the interface. To match the arithmetic mean (AM) with the integral formulation over a harmonic interpolation ansatz for C_V' in eq. (6.28), we introduce a correction factor f by

$$C_V'(\phi) = \frac{f C_V^\alpha C_V^\beta}{C_V^\alpha + \left(C_V^\beta - C_V^\alpha\right) h_\alpha(\phi)}. \tag{6.29}$$

A physical interpretation for our proposition is that while we locally violate the balance laws for the internal energy the global average across the interface is maintained, thus causing no eventual error to our energy balance. Inserting expression (6.29) in eq. (6.28) and replacing dx by $\frac{dx}{d\phi}d\phi$, we get for a binary system

$$\frac{\lambda}{2}\left[C_\alpha + C_\beta\right] = \int_0^1 \frac{f C_V^\alpha C_V^\beta}{C_V^\alpha + \left(C_V^\beta - C_V^\alpha\right) h_\alpha(\phi)} \frac{dx}{d\phi}\,d\phi. \tag{6.30}$$

With regards to the steady state solution for the evolution equation of ϕ [154], we can use the identities $dx/d\phi = (\pi\epsilon/4)\left(\sqrt{\phi(1-\phi)}\right)^{-1}$ and $\lambda = \pi^2\epsilon/4$, which leads to

$$\frac{\pi}{2}\left[C_V^\alpha + C_V^\beta\right] = \int_0^1 \frac{f C_V^\alpha C_V^\beta}{C_V^\alpha + \left(C_V^\beta - C_V^\alpha\right) h(\phi_\alpha)} \frac{d\phi}{\sqrt{\phi(1-\phi)}}. \tag{6.31}$$

This equation can only be solved numerically for f with respect to certain values of C_V^α and C_V^β. Before doing this we consider two additional possible factorisations: i) contrary to applying a constant f throughout

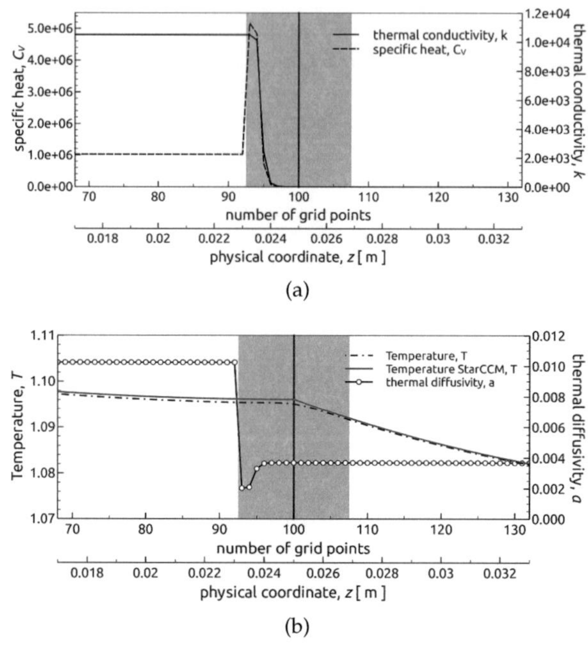

(a)

(b)

Figure 6.11: Inverse/harmonic interpolation in k and factorised inverse/harmonic interpolation of C_V using a phase dependent decaying factor. (a) interface profile ϕ together with material properties k and C_V, (b) temperature profile, interface profile ϕ and thermal diffusivity a. The shadowy area indicates the diffuse interface region, with a vertical line at $\phi = 0.5$.

Table 6.1: Numerical estimates on three different types of the factorised inverse interpolation of $C_V(\phi)$.

description	factor f
single fixed factor, f	$f = 3.81468$
single decaying factor, $fh(\phi_\alpha)$	$f = 3.82721$
phase dependent decaying factor, $(fh(\phi_\alpha) - h(\phi_\alpha) + 1)\,C_V^\alpha$	$f = 5.05906$

the whole interface, we assume f to be a function of ϕ_α, i.e. $f(\phi_\alpha) = fh(\phi_\alpha)$, and ii) we impose a factor only on one phase, e.g. in the phase

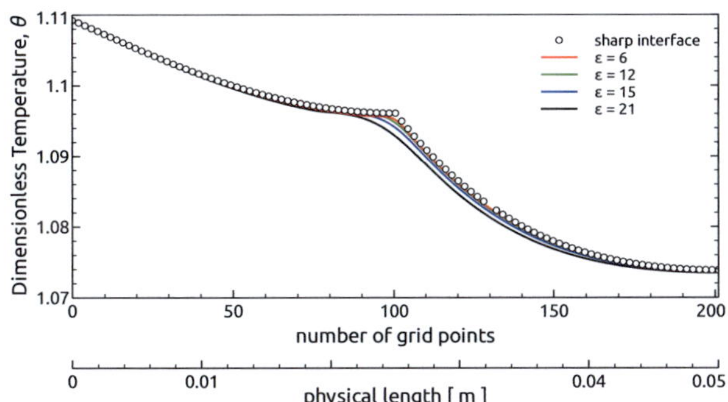

Figure 6.12: Interpolation of material properties k and C_V, interpolation of thermal diffusivity a and temperature profile, using decaying factorised inverse/harmonic interpolation in C_V for different diffuse interface widths, ϵ.

with the larger C_V, by $\tilde{C}_V^\alpha = (f h(\phi_\alpha) - h(\phi_\alpha) + 1) C_V^\alpha$ inside the interface $0 < \phi < 1$. Hence, eq. (6.31) can be rewritten to

$$\frac{\pi}{2} \left[C_V^\alpha + C_V^\beta \right] = \int_0^1 \frac{\tilde{C}_V^\alpha C_V^\beta}{\tilde{C}_V^\alpha + \left(C_V^\beta - \tilde{C}_V^\alpha \right) h(\phi_\alpha)} \frac{d\phi}{\sqrt{\phi \, (1 - \phi)}}. \tag{6.32}$$

For all the factorisations, we maintain the bulk values C_V^α and C_V^β outside the interface, respectively. Using $h(\phi) = \phi^3 (6\phi^2 - 15\phi + 10)$ as interface interpolation function, $C_V^\alpha = 64015.434$ and $C_V^\beta = 31.600714$ the numerical solution, using e.g. Mathematica, provides the factors listed in tab. 6.1.

Analogous to the representations in the previous sections, the distributions of the order parameter, the interpolated material parameters and the temperature are depicted in figs. 6.9, 6.10 and 6.11 for the single fixed factor, single decaying factor and the phase dependent decaying factor respectively.

All three factorisations show a steep increase in C_V with different magnitudes at the transition from bulk to the interface regarding the phase with the higher C_V value. The differences obtained for the different factorisations are more pronounced for considering the thermal diffusivity a. While the amplification, evoked by the fixed factor, keeps

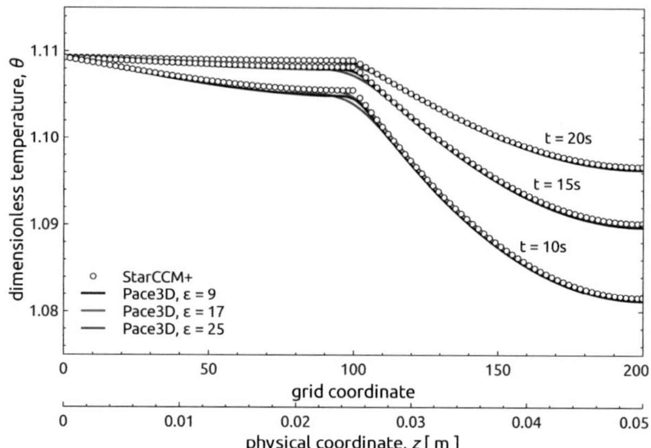

Figure 6.13: Interpolation of material properties k and C_V, interpolation of thermal diffusivity a and temperature profile, using decaying factorised inverse/harmonic interpolation in C_V for different diffuse interface widths, ϵ and at different instances of time.

the thermal diffusivity inside the interface at an almost constant level, it is gradually smoothed towards the bulk value of the second phase for the single decaying factor, cf. 6.9(b) and 6.10(b).

Due to the lump amplification in case of the fixed factorisation, the temperature inside the second phase (grid coordinate > 100) is slightly smaller than that of the reference solution. The matching of the temperature distribution is further improved using the decaying factorisation and finally the phase dependent factorisation, cf. 6.10(b) and 6.11(b) respectively.

Figure 6.12 shows the dependence of the temperature distribution for different diffuse interface widths. The maximum difference between the evaluated temperature curve and the reference solution is of an order of about 1%. We note that in the case $\epsilon = 21$, the interface consists of more than 50 cells, i.e. more than 25% of the computational domain. This high resolution is only chosen to elaborate the convergence properties of ϵ. In general when modelling in the context of phase-field methods, the interface is intended to be small compared to the overall computational domain.

Finally we proof the validity of the model for different instances of time, namely 10, 15 and 20s. Figure 6.13 shows a good overall matching of the temperature distributions for three different interface widths and for the three instances of time with the reference solution respectively.

In summary the effects of the different types of interpolation in terms of the temporal temperature difference and the temperature profile at an arbitrary timestep in fig. 6.14. Despite, that it correctly renders the effective thermal diffusivity, the linear interpolation causes a jump in temporal temperature difference as well as in temperature profile, cf. fig. 6.5. On the other hand, the inverse interpolation gives a smooth distribution of temporal temperature difference and temperature profile, whereas by definition it cannot reproduce the effective capacity – thus, a higher thermal diffusivity results in a higher temporal temperature difference and an increased temperature profile, cf. fig. 6.6. By design, the factorised inverse interpolation renders the correct effective capacity in the interface, as well as the corresponding thermal diffusivity, and finally avoids a jump across the interface.

6.5 Tensorial formulation

Based on the feasible model for one dimensional heat conduction perpendicular to the interface, we will extent the formulation to three dimensions. In the subsequent description, we split up the temperature diffusion equation

$$C_V \, \partial_t T = \langle \nabla, k \nabla T \rangle \tag{6.33}$$

in a flux-part

$$q = k \, \nabla T \tag{6.34}$$

and a divergence part

$$C_V \, \partial_t T = \langle \nabla, q \rangle . \tag{6.35}$$

With respect to the derivations in section 6.4.2, 6.4.3 and according to [158], we replace the scalar thermal conductivity k by a tensorial mobility $\mathbf{K} = \mathbf{K}(\phi, \nabla \phi)$, which depends on the direction relative to the interface

$$\mathbf{K} = k_\perp \mathbf{Q} + k_\parallel (\mathbb{1} - \mathbf{Q}), \tag{6.36}$$

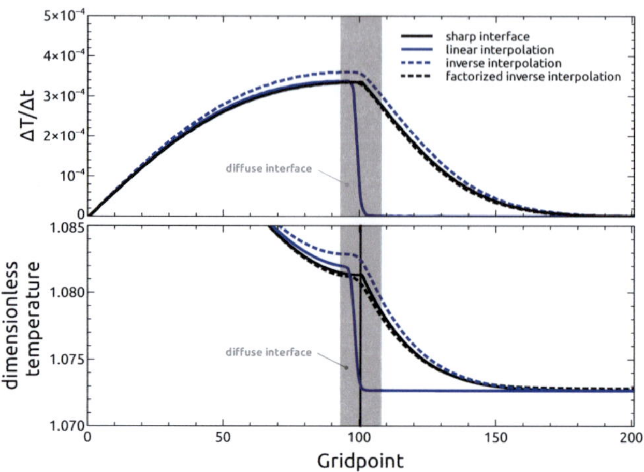

Figure 6.14: Temporal temperature difference and the dimensionless temperature profiles for a 1D simulation of a equally spaced composite domain (aluminium/air) heated from one side, for the linear interpolation, the inverse interpolation and the factorised inverse interpolation of $C_V(\phi)$ in the diffuse interface compared to the sharp interface solution.

where $\mathbb{1}$ represents the unit tensor, and $k_\parallel = k_\parallel(\phi)$ and $k_\perp = k_\perp(\phi)$ are the interpolation functions for heat flux parallel and perpendicular to the interface,

$$k_\parallel(\phi) = \sum_\alpha h(\phi_\alpha)k_\alpha \qquad \text{and} \qquad \frac{1}{k_\perp(\phi)} = \sum_\alpha \frac{h(\phi_\alpha)}{k_\alpha} \qquad (6.37)$$

respectively. The tensor $\mathbf{Q} = \mathbf{Q}(\nabla\phi)$ is defined as

$$\mathbf{Q} = \frac{\nabla\phi}{|\nabla\phi|} \otimes \frac{\nabla\phi}{|\nabla\phi|}. \qquad (6.38)$$

Using this, we write the flux part of the temperature diffusion equation as

$$q = \mathbf{K}\nabla T. \qquad (6.39)$$

With respect to the electric analogy outlined in sec. 6.3, we have a *circuit* for the heat current normal to the interface and two *circuits* for the heat currents in tangential directions, which obey different interpolation schemes. The actual heat conduction problem through the interface thus is a super-position of two parallel and one series *circuits*, i.e. the resulting evolution equation for the temperature must read

$$\partial_t T = \frac{1}{C_{V,\perp}}\partial_n q_n + \frac{1}{C_{V,\parallel}}\partial_s q_s + \frac{1}{C_{V,\parallel}}\partial_t q_t \tag{6.40}$$

where n, t, s are the coordinates normal and tangential to the interface, respectively.

Accordingly, we define a modified divergence operator in Cartesian co-ordinates as

$$\widetilde{\boldsymbol{\nabla}}_{\mathcal{R}} = \left(\mathcal{R}\,\mathbf{Q} + (\mathbb{1} - \mathbf{Q})\right)\nabla, \tag{6.41}$$

where $\mathcal{R} = C_{V,\parallel}/C_{V,\perp}$ is the ratio of the interpolation functions for the specific heat parallel and perpendicular to the interface,

$$C_{V,\parallel} = \sum_{\alpha} h(\phi_\alpha) C_V^\alpha, \tag{6.42}$$

$$C_{V,\perp} = \frac{\widetilde{C}_V^\alpha C_V^\beta}{\widetilde{C}_V^\alpha + \left(C_V^\beta - \widetilde{C}_V^\alpha\right) h_\alpha(\phi)}, \tag{6.43}$$

the latter being the decaying phase dependent factorised harmonic interpolation for a binary system, cf. dec. 6.4.6. We now rewrite the evolution equation as

$$C_{V,\parallel}\,\partial_t T = \langle\widetilde{\boldsymbol{\nabla}}_{\mathcal{R}}, \mathbf{K}\nabla T\rangle. \tag{6.44}$$

Note, that in the bulk regions, where $\mathcal{R} = 1$, we retrieve the classical divergence operator.

Equation (6.44) is solved in **PACE3D** on a structured, equally spaced Cartesian grid, by means of an explicit finite difference formulation. At this point, it is crucial to note that the formalism in [168] is based on a steady state solution, whereas the accuracy and stability of a numerical time dependent solution will also depend on the choice of the discrete time step. In the course of model development we are using a rather simple fully explicit finite difference framework, employing fully explicit forward Euler time discretisation. The three dimensional spatial implementation is done by means of a MAC (marker and cell) scheme with a staggered grid arrangement with respect to the three normal components and three shear components of the symmetric tensors $\mathbf{K}$. Therefore, timestep size is chosen with respect to the well known stability and quality criterions given in the literature [101, 102].

6.6 Validation

6.6.1 One dimensional semi infinite contact problem

In addition to the previous numerical survey, we continue with a one dimensional validation for the heat flux perpendicular to the interface, for two semi-infinite phases α and β in contact, having different material properties k^α, c_V^α and k^β, c_V^β as well as different initial temperatures T_0^α and T_0^β at time $t = 0$, cf. fig. 6.15. The interface position is assumed to be at the coordinate $x = 0$.

For each of the phases, we consider a one-dimensional heat conduction equation

$$C_V^\alpha \partial_t T^\alpha = \partial_x(k^\alpha \partial_x T^\alpha), \; x \leq 0 \tag{6.45}$$

$$C_V^\beta \partial_t T^\beta = \partial_x(k^\beta \partial_x T^\beta), \; x \geq 0. \tag{6.46}$$

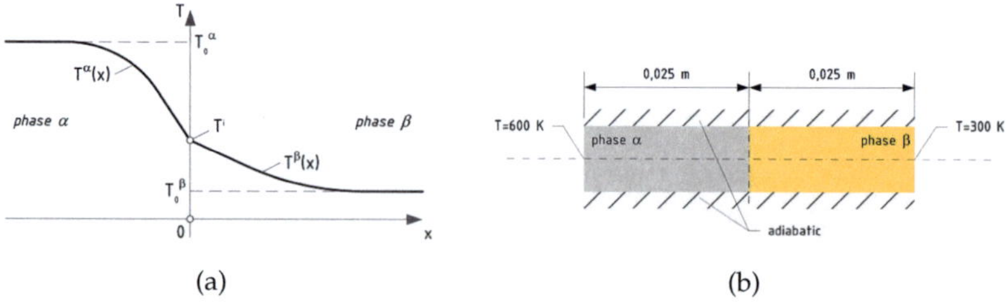

Figure 6.15: (a) One dimensional validation of heat conduction for two semi infinite phases α and β, with initial temperatures T_0^α and T_0^β respectively, and (b) setup for the numerical test case.

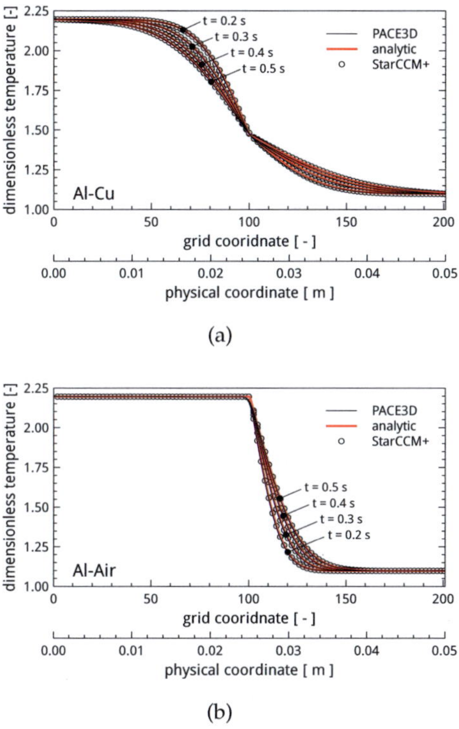

Figure 6.16: One dimensional heat conduction perpendicular to interface for (a) aluminium-copper and (b) aluminium-air for different instances of time.

In the context of a sharp interface representation the two equations are coupled by the interfacial temperature T^i given by

$$T^\alpha\big|_{-0} = T^\beta\big|_{+0} = T^i \tag{6.47}$$

and by the continuity condition of the heat flux

$$k_\alpha \partial_x T^\alpha\big|_{-0} = k_\beta \partial_x T^\beta\big|_{+0}. \tag{6.48}$$

As indicated in the textbooks [9, 167], the interfacial temperature T^i does not depend on time. Almost instantly after contact the interfacial

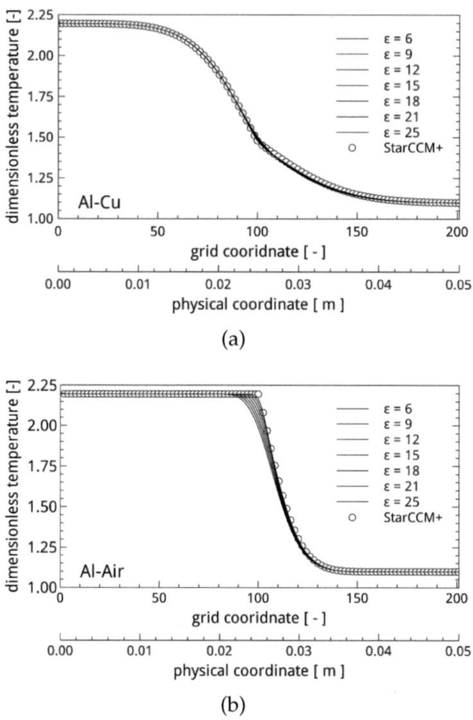

Figure 6.17: 1D heat conduction perpendicular to the interface for for different widths of the diffuse interface for (a) aluminium-copper and (b) aluminium-air.

temperature T^i establishes, and the temperature distribution inside the phases develops, depending on the interfacial temperature T^i and the constant initial temperature T_0^α and T_0^β. From [9], the interfacial temperature can be expressed as

$$T^i = T_0^\alpha + \frac{I^\alpha}{I^\alpha + I^\beta}(T_0^\beta - T_0^\alpha),\qquad(6.49)$$

where I^α and I^β are the thermal inertias of phases α and β respectively. Using the definitions $\xi^\alpha = x/\sqrt{4\,a^\alpha t}$ and $\xi^\beta = x/\sqrt{4\,a^\beta t}$, where

a is the thermal diffusivity, the temporal evolution of the temperature distribution is given as:

$$T^{\alpha}(x,t) = T_0^{\alpha} + (T^i - T_0^{\alpha})\,\mathrm{erfc}(-\zeta_{\alpha}),\ x \leq 0 \tag{6.50}$$

$$T^{\beta}(x,t) = T^i + (T_0^{\beta} - T^i)\,\mathrm{erf}(\zeta_{\beta}),\ x \geq 0. \tag{6.51}$$

For the purpose of validation we compare analytical as well as numerical calculations utilising the software PACE3D and StarCCM. We employ two different ratios for the material properties: the material combination aluminium-copper and aluminium-air.

For both composites and for different physical instances of time, $t = 0.2\,\mathrm{s}$, $0.3\,\mathrm{s}$, $0.4\,\mathrm{s}$ and $0.5\,\mathrm{s}$, we get an excellent agreement of the numerical and analytical solutions, as depicted in fig. 6.16. For the results obtained with PACE3D, a typical setting of $\varepsilon = 6$ is used for the diffuse interface width, which results in approximately 15 cells spanning the diffuse interface.

In order to depict the influence of the interface width, fig. 6.17 shows the results of different settings for ε for both combination of materials at a fixed time $t = 0.3\,\mathrm{s}$. Due to the high ratios in k and C_V, the combination of aluminium and air shows slight variations of the temperature distributions inside the diffuse interface as ε varies, whereas the bulk distributions match. When material properties are of the same order of magnitude, e.g. for the combination of aluminium and copper, the effect is less significant. Despite of the variations inside the diffuse interface, which are process related, the overall temporal and spatial distribution of temperature is met in good accordance with the analytical and numerical data.

6.6.2 Two dimensional circular inclusion

Validation of two dimensional transient tensorial heat conduction is done with a simple geometric setup. We consider a rectangular domain of $0.02\,\mathrm{m} \times 0.02\,\mathrm{m}$ that contains a circular disc with a diameter of $0.008\,\mathrm{m}$. The domain is discretised on a uniform rectangular grid of 200×200 cells.

While we apply properties of aluminium for the circular body, we use different material combinations for the enclosing rectangular domain in

Table 6.2: Thermal properties of materials used in the simulations. ρ: density $[\mathrm{kg\,m^{-1}}]$, k: thermal conductivity $[\mathrm{W\,m^{-1}\,K^{-1}}]$, c_V: specific heat $[\mathrm{J\,kg^{-1}\,K^{-1}}]$.

phase	abbrev.	ρ	k	c_V
Copper	Cu	8940	398	836
Aluminium	Al	2800	150	837
Steel	Fe	7870	80	448
Stainless Steel	CrNiMo	7900	17	460
Polytetrafluorethylen	PTFE	2200	0.2	1000
Air	Air	1.15	0.025	1006

order to validate our model for different ratios of the thermal conductivity and specific heat. The properties for the various composites are given in tables 6.2 and 6.3.

As initial condition we impose a temperature of 600 K on the disc, while the surrounding domain is set to 300 K. For all boundaries of the rectangular domain we set adiabatic conditions, i.e. $\partial_n T = 0$.

We report numerical calculations using the commercial package Star-CCM and PACE3D in comparison for all composites listed in tab. 6.3. The PACE3D simulations are performed using a diffuse interface width of about 15 cells (according to a model parameter for the interface width of $\varepsilon = 6$). Evaluation of the temperature distribution along a horizontal cross section at different instances of time is shown in fig. 6.18 for all composites. The comparison gives a good overall compliance for all composites, i.e. for ratios of the thermal conductivity k (tab. 6.3) ranging from ≈ 1 up to ≈ 6000 and of the volumetric heat capacity C_V ranging

Table 6.3: Composites and their ratios of the thermal conductivities k and volumetric heat capacities C_V.

composites	abbrev.	ratio of k's	ratio of C_V's
Aluminium-Steel	Al-Fe	~ 1	~ 1
Aluminium-Stainless Steel	Al-CrNiMo	~ 2	~ 1
Aluminium-Polytetrafluorethylen	Al-PTFE	~ 107	~ 1
Aluminium-Air	Al-Air	~ 6000	~ 2026

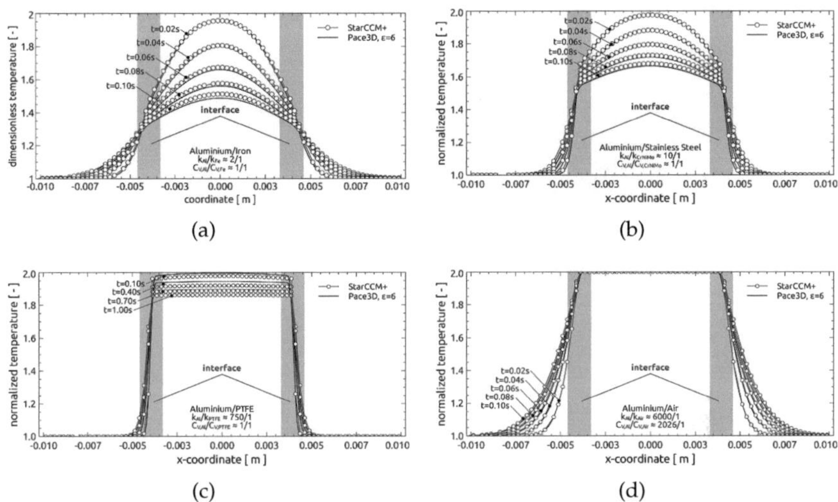

Figure 6.18: Comparison of the temperature distributions for a two dimensional circular inclusion at a horizontal cross section obtained with PACE3D and Star-CCM for different instances of time and for material combinations (a) aluminium/iron, (b) aluminium/stainless steel, (c) aluminium/PTFE, (d) aluminium/air. The shadowy regions mark the diffuse interface layer.

from ≈ 1 up to ≈ 2025. A maximum deviation between the solutions of about 1 % is found, which might even be traced back to the different numerical schemes of the software packages.

6.6.3 Three dimensional spherical inclusion

Finally we perform a full three dimensional calculation to test the tensorial mobility formulation. We use a spherical inclusion with a diameter of 0.008 m in a cubic domain with an edge length of 0.016 m and a computational representation of 80 cells in each direction.

We impose the same temperature conditions for the body and boundaries of the domain as in 2D. As composite we apply aluminium/stainless steel and compare the results of PACE3D with the reference solution StarCCM.

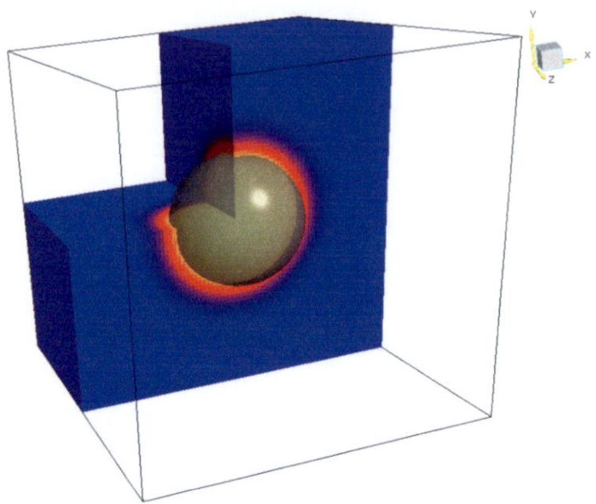

Figure 6.19: 3D view of the three dimensional spherical inclusion. The transparent sphere indicates the spatial dimensions, while the volume shape is coloured using the temperature distribution.

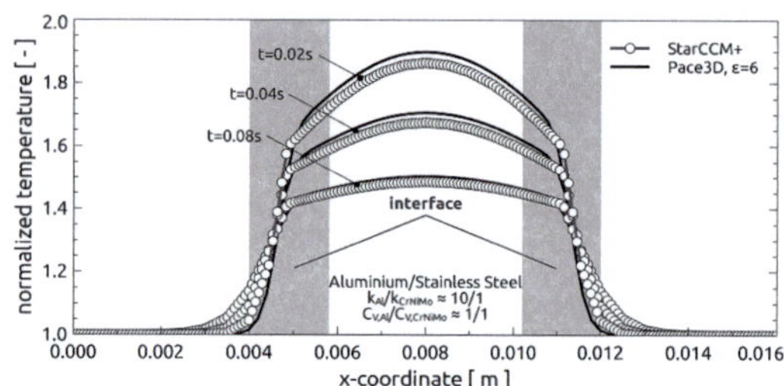

Figure 6.20: Comparison of temperature distributions obtained with PACE3D and StarCCM at a central cross section of the sphere for different instances of time.

Figure 6.20 contains the temperature distributions at a central cross section of the sphere at different times. Results are in good agreement with the reference solution, even though we observe a slightly increased difference, which can be explained by the lower resolution (lesser number of grid points) used in 3D. During the evolution in time, the deviation reduces.

6.7 Segmented tensorial model

Regretfully, we still have to struggle with the application of the above mentioned method in case of high ratios in material properties for complex geometries. This is the motivation to further investigate these deficiencies, and an improvement is presented hereby. For comprehension, it is useful to recapitulate the development of the method in the following.

6.7.1 Outline

Diffuse Interface transport problem

Consider the general transient two-phase transport processes, with static interfaces. For the phases α and β, we write

$$C_i \, \partial_t \theta = \nabla \cdot (M_i \, \nabla \theta) \qquad \text{with} \qquad i \in \alpha, \beta. \tag{6.52}$$

Here, θ is the potential, the M_i's and C_i's are the phase dependent mobilities and capacities. At the interface the continuity of the potential and of the flux holds

$$\theta_\alpha = \theta_\beta \qquad \text{and} \qquad M_\alpha \nabla \theta \big|_\alpha = M_\beta \nabla \theta \big|_\beta . \tag{6.53}$$

In the phase-field context just one diffusion equation is considered

$$C(\phi) \, \partial_t \theta = \nabla \cdot (M(\phi) \, \nabla \theta) , \tag{6.54}$$

where $M(\phi)$ and $C(\phi)$ describe the transition of the bulk mobilities and capacities across the interface, and ϕ represents the order parameter in the context of phase-field model.

Tensorial mobility approach

Originally, the mobility M and the capacity C are scalar quantities. As outlined above, a scalar formulation of the governing equations in the phase-field context induces so called diffuse-interface effects: *surface-diffusion* will lead to artificial surface currents along the diffuse interface, *surface-resistance* will cause discontinuities in the bulk potential extrapolated to the interface point of symmetry, at $\phi = 0.5$ representing the sharp-interface limit.

In [158] it is shown, that both effects can be eliminated simultaneously, by introducing the tensorial mobility approach. He focused on steady state transport processes

$$\nabla \cdot (\mathbf{M}(\phi, \nabla\phi)\, \nabla\theta) = 0, \tag{6.55}$$

from which he derived the tensorial formulation and the interpolation of the mobilities herein, with respect to the interface orientation:

$$\mathbf{M}(\phi, \nabla\phi) = M^{\perp}(\phi)\mathbf{Q} + M^{\|}(\phi)(\mathbb{1} - \mathbf{Q}), \tag{6.56}$$

where $\mathbb{1}$ represents the unit tensor, and $\mathbf{Q}$ is the projection onto the interface normal $\mathbf{n} = \frac{\nabla\phi}{|\nabla\phi|}$, given by

$$\mathbf{Q} = \mathbf{n} \otimes \mathbf{n}. \tag{6.57}$$

The different interpolation functions for the mobilities perpendicular and parallel to the interface, are $M^{\|}(\phi) = M_\alpha \phi_\alpha + M_\beta \phi_\beta$ and $M^{\perp}(\phi) = (M_\alpha^{-1}\phi_\alpha + M_\beta^{-1}\phi_\beta)^{-1}$.

Modified divergence operator

An extension of this approach to three dimensions and transient transport problems is given above, where a modified divergence operator is derived. With dedicated interpolation functions, the capacities parallel and perpendicular to the interface are $C_{V,\|}(\phi)$ and $C_{V,\perp}(\phi)$ according to eqns. (6.42) and (6.43), respectively. The heat conduction equation is rewritten as

$$C_{\|}(\phi)\,\partial_t\theta = \langle \widetilde{\nabla}_{\mathcal{R}}, \mathbf{M}(\phi, \nabla\phi)\nabla\theta \rangle. \tag{6.58}$$

Herein, $\widetilde{\nabla}_{\mathcal{R}}$ represents a spatially varying divergence operator given in Cartesian coordinates, locally rotated with respect to the interface orientation and weighted according to the directional interpolation scheme of $C_{V,\perp}$ and $C_{V,\|}$, respectively. If written in interface normal coordinates, with orthonormal basis vectors $\mathbf{n}$, $\mathbf{s}$ and $\mathbf{t}$, the operator reads as

$$\widetilde{\nabla}_{\mathcal{R},(n,t,s)} = \mathcal{M}_{\mathcal{R},(n,t,s)} \nabla_{(n,t,s)} = \begin{pmatrix} \mathcal{R} & 0 & 0 \\ 0 & 1 & 0 \\ 0 & 0 & 1 \end{pmatrix} \nabla_{(n,t,s)} \tag{6.59}$$

In the same sense, one could think of $\mathbf{M}\nabla$ as a spatially varying gradient operator given in Cartesian coordinates, locally rotated with respect to the interface orientation as eq. (6.59) and weighted according to the directional interpolation schemes of $M^{\perp}$ and $M^{\parallel}$, respectively.

6.7.2 Discussion

The approach of [158] limits the applicability to steady two-dimensional transport processes, and simulation results where shown for maximum ratios in mobilities up to 10. In the preceding sections, the extension to a more general transient and three-dimensional model was realised, which also proofed to be feasible for slightly higher ratios in mobilities and capacities.

Both models provide a solid foundation for the correct description of the physical scope of interest. The persistent limitation for both models, according to high distinctive ratios in material properties and complex geometries, is a matter of the implementation, done by means of a MAC (marker and cell) scheme. Whereas the mapping of the mobilities and the discretisation of the derivatives was done as described in [158], the conductivities in sec. 6.5 were subject to central discretisation schemes.

In the vicinity of high curvatures and steep gradients in mobility, capacity and the order parameter, the method resulted in checkerboard-like solutions and growing oscillations of the potential. Reduction of time step width, as well as different staggered discretisation approaches, did not lead to success. Hence, a so called *segmented* approach is considered, and presented below.

6.7.3 Formulation

In order to identify the sources of instabilities, we separate eq. (6.58) by applying the product rule, and rearrange the resulting terms by the order of derivatives of potential

$$C_{V,\parallel}\,\partial_t\theta = \langle \widetilde{\boldsymbol{\nabla}}_{\mathcal{R}}, \mathbf{M}\nabla\theta \rangle = \langle \mathcal{M}_{\mathcal{R}}\nabla, \mathbf{M}\nabla\theta \rangle \tag{6.60}$$
$$= \langle \mathcal{Q}\nabla, \nabla\theta \rangle + \langle \mathcal{L}, \nabla\theta \rangle,$$

where $\widetilde{\boldsymbol{\nabla}}_{\mathcal{R}}$ is the modified divergence operator with respect to Cartesian co-ordinates.

The coefficient matrix $\mathcal{Q} = \mathcal{M}_\mathcal{R}\mathbf{M}$ and the differential operator $\mathcal{Q}\nabla$ have the same form as eq. (6.59) in Cartesian co-ordinates. If this differential operator is applied on $\nabla\theta$, we will write this expression as the tensor contraction $(\mathcal{M}_\mathcal{R}\mathbf{M}) : \mathcal{H}(\theta)$, where $\mathcal{H}(\theta)$ is the Hessian of the potential. With respect to the order of the Hessian, we name it the quadratic part of the tensorial formulation.

Deriving the coefficient vector $\mathcal{L} = \widetilde{\boldsymbol{\nabla}}_\mathcal{R} \cdot \mathbf{M}$, we apply the modified divergence operator $\widetilde{\boldsymbol{\nabla}}_\mathcal{R}$ on $\mathbf{M}$. We then scalar multiply it with the Jacobian of the temperature, and thus, with respect to the order of $\nabla\theta$, we name it the linear part of the tensorial formulation.

We exclusively attribute diffusive character to the quadratic term $(\mathcal{M}_\mathcal{R}\mathbf{M}) : \mathcal{H}(\theta)$, since it just contains second order derivatives of θ and constant coefficients. With regard to the structure of a general advection-diffusion equation, and because of the stationary pseudo velocity $\mathcal{L}$, we consider the linear part on the right-hand side as a pseudo-advective term.

This kind of formulation reveals the source of instabilities. In case of high ratios of mobilities and high curvature in ϕ, the contribution of the linear part is getting significant. Owing to its advective nature, this contribution fails with a central discretisation scheme with increasing pseudo velocity, and promotes instabilities. Therefore, the instabilities we received by applying purely central schemes on the previous formulation, using the equivalent eq. (6.58), are now obvious. The staggered implementation showed reasonable and stable solutions for ratios in thermal conductivity up to about 500, while the curvature of the validation cases was moderate for the circular and spherical inclusions tested, cf. sec. 6.6.

The new formulation allows us to apply different discretisation schemes on the quadratic and linear part, respectively. According to the standard discretisation methods given in [101, 168, 200], we use a central discretisation scheme on the diffusive part and an upwind scheme for the pseudo-advective part, according to the quadratic and linear terms respectively.

6.8 Validation of the segmented approach

A comprehensive study and validations of different interpolation methods on one-dimensional test cases are already given in sec. 6.6. In the following, we provide numerical calculations in two dimensions, to incorporate the influence of interface orientation and thus of the different directional interpolations of the bulk properties. Accordingly, we assess the accuracy and the stability of the calculations for different setups.

Whereas the model and discretisation was given in terms of a general transient transport equation, transient heat conduction is a representative application. In the scope of this work we apply the corresponding transient temperature equation

$$C_{V,\|}\, \partial_t T = \langle \mathbf{M}_\mathcal{R} \mathbf{K} \nabla, \nabla T \rangle + \langle \tilde{\boldsymbol{\nabla}}_\mathcal{R} \cdot \mathbf{K}, \nabla T \rangle, \qquad (6.61)$$

and solely narrow down the used material-combination to aluminium and air, in order to reflect one of the most different material settings from an engineering point of view, cf. tab. 6.2. In this context, the potential θ, mobility M and capacity C correspond to the temperature T, thermal conductivity k and volumetric heat capacity C_V, respectively.

If not otherwise stated, a normalised quadratic domain is discretised on a uniform rectangular grid of 100×100 cells, corresponding to the physical lengths of $0.1\,\text{m} \times 0.1\,\text{m}$. While using a reference temperature of $T_{ref} = 300\,\text{K}$, a reference length of $l_{ref} = 0.1\,\text{m}$ and the thermal conductivity and volumetric heat capacity of air for normalisation, we keep the units for the time in the following figures to improve comprehensibility, cf. tab. 6.2.

We consider different setups, according to the direction of heat flux with respect to the interface. Heat conduction perpendicular to the interface is considered in a quadratic domain of normalised dimensions 1×1 with a quadrant of radius 0.7, placed at the lower left corner. The initial temperatures of aluminium and air are $T_{Al} = 2$ and $T_{Air} = 1$ respectively, while the domain boundaries are defined adiabatic.

In order to investigate the heat conduction parallel to the interface, we use a quadratic domain of dimensions 1×1, which is divided into two equal rectangular subdomains of dimensions 0.5×1. The whole domain is initialised with a temperature of $T_{Al} = T_{Air} = 1$. A constant temperature is applied at one boundary perpendicular to the interface, while all other boundaries are adiabatic.

From the first setup we derive two test cases, by applying the material properties of aluminium on the circular inclusion and the material properties of air on the surrounding, and vice versa. They are named *convex*

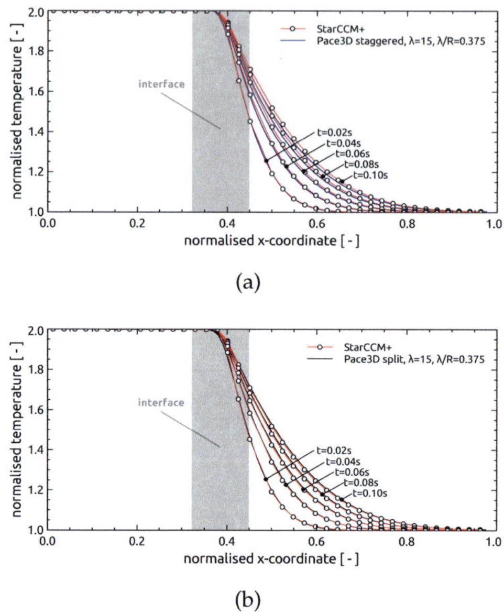

Figure 6.21: Temperature profiles across the convex interface (circular inclusion of aluminium in air) at different times for the setup in sec. 6.6.2, fig. 6.18(d), where a circular inclusion of radius 0.4 m is embedded symmetrically into a rectangular domain of 0.02 m × 0.02 m, resolved by 200 × 200 cells. The temperature distributions of the (a) staggered and the (b) segmented approach are compared to the numerical reference solution obtained with StarCCM. The shadowy regions mark the diffuse interface layer.

interface and *concave interface,* with respect to the curvature of the aluminium phase boundary. Evaluation is done on a radially symmetrical cross section of the inclusion. Finally, the case with equal rectangular subdomains is called *parallel interface,* where the temperature distributions are determined at a cross section parallel to the hot boundary.

6.8.1 2D circular inclusion – convex interface

At first we racapitulate the deficiency of the staggered implementation of [158] which was applied to a two-dimensional convex interface setup in sec. 6.6, cf. fig. 6.21(a). The temperature distributions obtained with PACE3D show deviations compared to the numerical reference solution of StarCCM, already for short simulation times. Since the errors in a diffuse transport process are accumulated over the time, the differences will become significant.

The results of the segmented-implementation (eq. (6.60)) on the same setup are in excellent agreement with the numerical reference solution of StarCCM, cf. fig. 6.21(b).

Aiming at simulating larger physical times, we performed long-time runs on the convex interface setup for 2 s and 10 s physical time. Figure 6.22 shows the temperature profiles for 2 s and 10 s respectively. We performed simulations with different proportions of interface width to

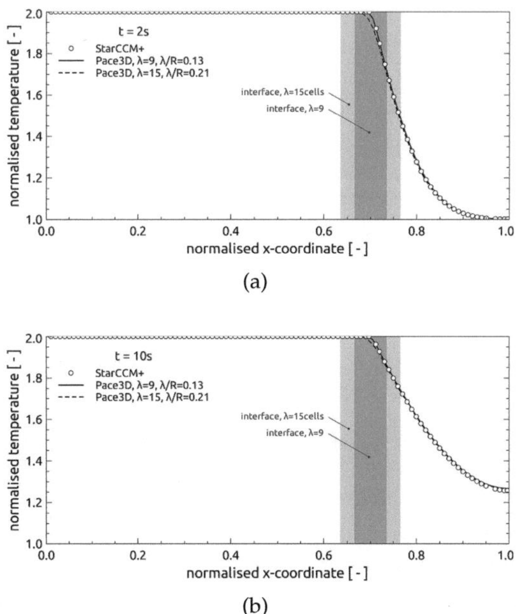

Figure 6.22: Temperature profiles across the convex interface (circular inclusion of aluminium in air) for long-time simulations (a) for 2 s and (b) for 10 s, using the segmented approach, each compared to the numerical reference solution obtained with StarCCM. The shadowy regions mark the diffuse interface layer.

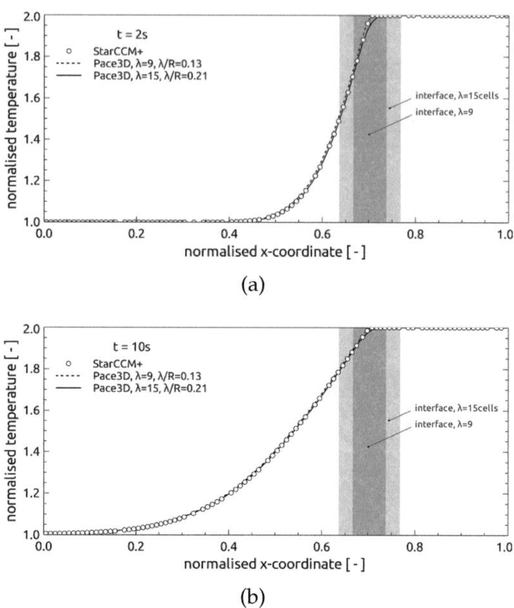

Figure 6.23: Temperature profiles across the concave interface (circular inclusion of air inside aluminium) for two different interface widths, 6 cells and 15 cells, respectively, after (a) two seconds and (b) ten seconds, each compared to the numerical reference solution obtained with StarCCM. The shadowy regions mark the diffuse interface layer.

radius, in particular $\lambda/R = 0.13$ and $\lambda/R = 0.21$. Both results are in good agreement with the numerical reference solution obtained with StarCCM. Minor deviations inside the diffuse interface region are intrinsic with respect to the diffuse interface modelling, compared to the sharp interface solution.

6.8.2 2D circular inclusion – concave interface

Interchanging the phases from the convex interface filling leads us to a concave interface setup. Figures 6.23(a) and 6.23(b) show the temperature distributions for the long-time runs for 2 s and 10 s physical time, respectively. The overall agreement with the numerical reference solution obtained with StarCCM is excellent. Note, that there is no significant difference between the solutions obtained for interface width ratios of $\lambda/R = 0.13$ and $\lambda/R = 0.21$.

6.8.3 2D parallel interface heat conduction

Using the parallel interface setup mentioned above, we apply material properties of air and aluminium for the left and the right subdomain, respectively. Both phases are subject to a constant temperature boundary condition at the bottom of the domain.

Note, there is no pure heat conduction parallel to the interface. Due to the difference in the material properties of the bulk phases, there is also a difference in the temporal evolution of temperature inside the two phases. As a consequence of this, a temperature gradient and a heat flux perpendicular to the interface will arise. However, this setup allows us to investigate the effect of heat conduction parallel to the interface.

Figure 6.24 shows the temperature distributions across the plain parallel heated interface at 20 % domain height (a) at $t = 2\,\mathrm{s}$ and (b) at $t = 10\,\mathrm{s}$. The solutions are obtained using different interface widths $\lambda = 6, 9$ and 12 cells, which has an almost vanishing effect. Generally speaking, the temperature distributions obtained with PACE3D, are in good agreement with the reference solution of StarCCM.

6.8.4 Consistency

Assessment of the segmented implementation is done by checking the conservation of the internal energy on the above-mentioned convex and concave-interface long-time simulations for $t = 10\,\mathrm{s}$. In the latter setups homogeneous Neumann conditions are used on all domain boundaries, thus the total internal energy

$$E^k = \sum_i \sum_\alpha \left[\phi_\alpha C_V^\alpha T_i \right] \Big|^k \, . \tag{6.62}$$

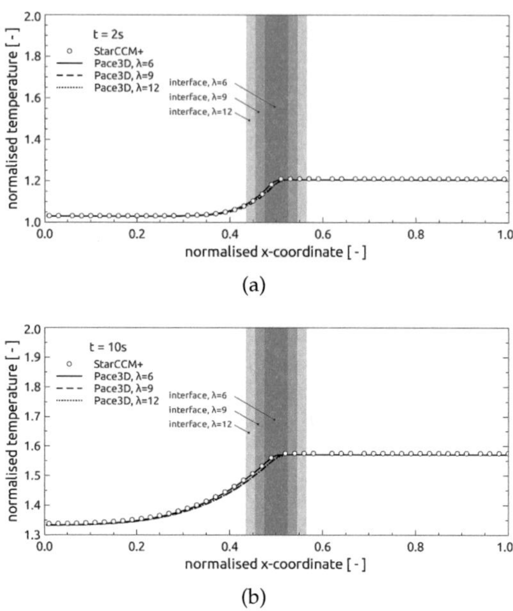

Figure 6.24: Temperature profiles across the parallel interface after (a) two seconds and (b) ten seconds, each compared to the numerical reference solution obtained with StarCCM. The shadowy region marks the diffuse interface layer.

must keep constant. The thermal energy at time step k is computed as sum over all grid points $i = 0, 1, \ldots, N$ and all phases α. On the ordinate we plot the relative energy error defined as

$$\epsilon^k_{energy} = \frac{E^k}{E^{k=0}} - 1,$$
(6.63)

where $E^{k=0}$ is the initial thermal energy at initial timestep $k = 0$.

From fig. 6.25 we find, that in each simulation considered, the relative error is less than 1.5 %. In [158] it is shown, that in the range of 0.005 $\leq \lambda/R \leq$ 0.035 the error vanishes for the steady formulation. We are aware of the fact, that due to the rather practical factorisation approach introduced for the transient formulation, we do not have an exact model up to now. Even though the perpendicular heat conduction matches

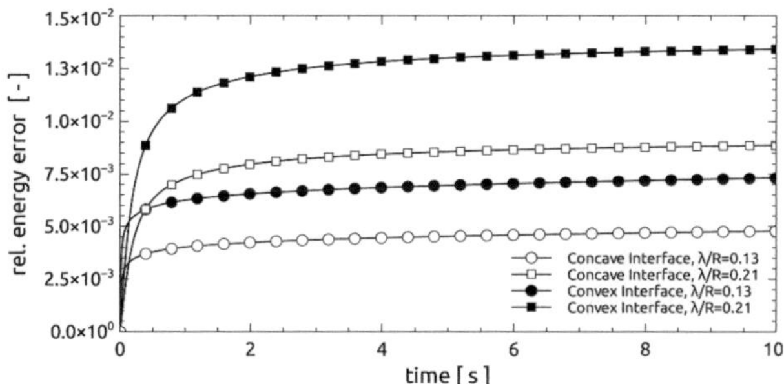

Figure 6.25: Relative error in overall thermal energy for concave and convex interfaces, each with different interface widths ($\lambda = 9$ cells and $\lambda = 15$ cells).

excellently, we still notice slight differences in the parallel interface setup, which is due to the missing dependency of the factorisation $f_{\alpha,\beta}$ on the interface width λ. Additionally, compared to [158], we have used much higher ratios λ/R, since we are interested on further applications with moderate grid resolutions.

Due to the intrinsic misalignment of phase boundaries and the numerical grid, the method is affected by the grid-anisotropy. This explains the difference of the errors of convex and concave setups. Furthermore, we notice that the absolute level of error is decreasing with the ratio λ/R, and the magnitude over time is bounded.

6.8.5 Effect of initial solution

For the simple two-dimensional validations, we encountered very small over-shoots ($\ll 1\,\%$) in temperature in the vicinity of the interface during the first few iterations, that vanished immediately. Owing to the sharp temperature distribution used as initial condition applied on a smooth interface, the temperature in the interface first has to settle. We performed numerical tests, using a smooth initial temperature distribution, that was scaled according to the order parameter, and successfully minimised the over-shoots. However, even though a scaled initial distribution behaves smoother than a sharp initial solution, in setups where two bodies of different temperature touch immediately at the first iteration, it is physically no better either. The latter is rather an analytical or numerical ideal conception, which induces the mentioned over-shoots.

6.9 Convective heat transfer

Besides the diffusive heat transport due to conduction, we have to consider the transportation of heat by the macroscopic convective movement in the case of coupled heat transfer and fluid flow. In this context, the term conjugate heat transfer [73] is used, and includes the temperature evolution in solids and fluids caused by the thermal and hydraulic interaction. Here, the contribution of diffuse or convective transport depends on the thermal properties of the materials (solid and fluid) as well as the flow regime. The latter, in turn, heavily depends on the geometry.

For the application of quasi-incompressible low speed flows (Ma $\ll$ 0.3), the viscous effects can be neglected as well as the pressure work term. Accounting for these prerequisites, the transient convection-diffusion equation for the temperature field reads

$$\underbrace{C_{V,\parallel}\,\partial_t T}_{\text{local derivative}} + \underbrace{\langle \boldsymbol{u}, \nabla T \rangle}_{\text{convective derivative}} = \underbrace{\langle \mathbf{M}_{\mathcal{R}} \mathbf{K} \nabla, \nabla T \rangle + \langle \widetilde{\boldsymbol{\nabla}}_{\mathcal{R}} \cdot \mathbf{K}, \nabla T \rangle}_{\text{diffusion (segmented approach)}},$$

$$(6.64)$$

where $\boldsymbol{u}$ is the fluid velocity, and the convective derivative contributes for the transportation of energy [6, 101, 102]. Note, that the local derivative $\partial_t T$ is the local time rate of change of the temperature at a fixed spatial location, whereas $\boldsymbol{u} \cdot \nabla T$ is the time rate of change due to the movement of the fluid. Both can be summarised as substantial derivative $D/Dt \equiv \partial_t + \boldsymbol{u} \cdot \nabla$ that can be applied on any flow field variable [6].

For technical processes, the convective heat transfer is usually characterised by the Nusselt number, which is introduced in the experimental part of this work, cf. eq. (3.9) in sec. 3.7.2.

6.10 Parametrisation

With respect to the computational treatment and accuracy we use a non-dimensional representation of the governing equations. Selecting basic reference quantities, all physical quantities (p) can be normalised accordingly; cf. parametrisation for the diffuse interface fluid flow in sec. 5.9. Using characteristic values (ref) for the reference length x_{ref}, the density ρ_{ref}, the time t_{ref} and the temperature T_{ref}, we write the dimensionless quantities (n) as

$$T_{(\mathrm{n})} = \frac{T_{(\mathrm{p})}}{T_{\mathrm{ref}}}, \quad x_{(\mathrm{n})} = \frac{x_{(\mathrm{p})}}{x_{\mathrm{ref}}}, \quad t_{(\mathrm{n})} = \frac{t_{(\mathrm{p})}}{t_{\mathrm{ref}}}, \quad \text{and} \quad \rho_{(\mathrm{n})} = \frac{\rho_{(\mathrm{p})}}{\rho_{\mathrm{ref}}}. \quad (6.65)$$

Accordingly, the physical values of specific heat capacity $c_{V,(\mathrm{p})}$, the volumetric heat capacity $C_{V,(\mathrm{p})}$ and the thermal conductivity $k_{(\mathrm{p})}$ are transferred into non-dimensional (n) values as

$$c_{V,(\mathrm{n})} = c_{V,(\mathrm{p})} \cdot \frac{t_{\mathrm{ref}}^2 T_{\mathrm{ref}}}{x_{\mathrm{ref}}^2}$$

$$C_{V,(\mathrm{n})} = C_{V,(\mathrm{p})} \cdot \frac{t_{\mathrm{ref}}^2 T_{\mathrm{ref}}}{\rho_{\mathrm{ref}} x_{\mathrm{ref}}^2} \qquad (6.66)$$

$$k_{(\mathrm{n})} = k_{(\mathrm{p})} \cdot \frac{t_{\mathrm{ref}}^3 T_{\mathrm{ref}}}{\rho_{\mathrm{ref}} x_{\mathrm{ref}}^4},$$

whereas a comprehensive and specific example is given appendix E.

Chapter 7

Application[1]

Up to here we have introduced the lattice-Boltzmann method for the modelling of fluid flow in chapter 5 and the evolution equation of temperature in chapter 6, both in the context of the phase field model, where special attention is payed upon the modelling in the diffuse interface region. In this chapter a brief outline of the numerical implementation is given, and the coupled diffuse fluid flow and diffuse heat transfer method is employed on the simulation of open cell metal foams.

With respect to the embarrassments that come along with the measurements of fluid flow and heat transfer at the pore scale level of open cell foams, it appears likely to apply numerical methods. However, simulation does not come at zero cost. The task is not only to provide a feasible geometrical and topological model for the foam under consideration, but also in the numerical treatment of the complex structure. Today, processing power is falling in cost, which allows for the simulation of even bigger domains and more complex physics. However, the pre- and post-processing tasks are still challenging, and the rule of thumb of the early times of numerical modelling – which states that about 80% of the effort is spent on mesh generation, model specification and evaluation of the results – still holds.

Even though sophisticated mesh generation tools are available in the field of academic as well as commercial simulation tools, the discretisation of a complex cellular solid is still a challenging, time consuming and most interactive task. To this effect, the combination of an automated method for foam generation (chap. 2), within the context of an interface capturing phase filed approach (chap. 4) with appropriate methods for fluid flow (chap. 5) and heat transfer (chap. 6) provides a promising alternative. Apart from that, to start with the geometry, a suitable representation of the foam structure is essential.

---

[1]Parts of the subsequent sections are submitted for publication in *Advanced Engineering Materials* [66] and *Cellular materials: Proceedings CELLMAT 2014* [63, 69].

To the authors best knowledge, there are only few publications on the coupled fluid flow and heat transfer simulation in open cell metal foams. Actually, the greatest difficulty is the realistic modelling of the foam structure. For instance, this can be realised by conducting expensive and time consuming investigations using industrial computed tomography (CT) scanning, which provides voxel data that requires careful post-treatment to prepare the three-dimensional foam structures, or applying a simplified modelling approach, cf. chap. 2. For the most part, validation of heat transfer and fluid flow performance is done against macroscopic parameters such as pressure drop, pressure drop per unit length and Nusselt number.

A numerical model for fluid flow modelling based on a fundamental periodic structure of eight unit cells is examined in the work of [19]. The foam geometry is spatially resolved by unstructured tetrahedral meshes with 2.2×10^5, 4.4×10^5 and 8.3×10^5 elements. Simulations are carried out using the flow solver **CFD-ACE** of **CFDRC**[2], whereas all pressure drop results are about 25% underestimated. The authors claim that responsibility for the discrepancy is due to the lack of side wall effects.

Numerical analysis of the conduction heat transfer in high porosity foam structures is done in [38]. The effective thermal conductivity of porous structures is most commonly modelled by means of empirical or semi-empirical models, based on different assumptions regarding pore scale and unit cell topology, which reveals significant variations. Therefore, a finite volume method is applied for evaluation and assessment. Foams are generated from regular structures (unit cells) as well as from tomographic data. The results show, that the fraction of solid phase in the struts and in the lumps (intersections) is the key parameter for successful modelling, whereas the shape of the cells and struts has much less impact.

A completely different approach is pursued in [123], where a so called mesh-based microstructure representation algorithm (MBMRA) is employed for the modelling of cellular solids. In doing so, a reasonable fine mesh spans the whole simulation domain including fluid and solid regions. Random placed seed points are used for a rule based unstructured mesh growing algorithm for the solid matrix. The residual cellular structure depends on the predefined rules, and facilitates the generation

---

[2]`http://www.cfdrc.com/` (accessed: 15/5/2014)

of porous or fibrous structures for instance. Finally, the mesh provides a sharp interface representation of the fluid and solid domain, whereas a coupled Navier-Stokes solver for the fluid and a heat conduction code for the solid is used to simulate the conjugate heat transfer problem in a porous structure. Despite the interesting MBMRA-approach, the presented results lack of a comprehensive validation.

In [12] a multiple relaxation time lattice Boltzmann method is employed to simulate the flow field in a metallic foam sample. The geometry of the sample is gained from CT-scans conducted for a single cubic NiCr-foam sample of about 12.8 mm edge length. Experimental pressure drop data is successfully recovered for a low to medium Reynolds number regime.

Foam structures based on a unit cell modelling are employed in [114] for the coupled simulation of fluid flow and heat transfer using an in-house solver for the Navier-Stokes and energy equations. Foam structures of 10 ppi and 40 ppi are generated from regular unit cells. Different unstructured computational grids with about 7×10^5, 1.5×10^6 and 2.8×10^6 tetrahedral elements were used, whereas the computational domain spans several pores in streamwise direction, in order to be representative. Pressure loss as well as heat transfer data is found to be reasonable with respect to experimental data. However, the authors in [114] consider, that the geometry and foam creation process is tedious and time consuming.

A commercial Navier-Stokes CFD-Software (Ansys FLUENT[3]) is used in [46] for the simulation of pressure drop in foam structures of different pore densities but similar relative porosity. Again, these authors received two-dimensional images through computer tomographic X-ray measurements, which are then used to derive three-dimensional foam structures as a starting point for flow simulations. Computational domains of different sizes, covering a representative section of the foam with respect to the number of pores, are generated for the samples with pore densities 5 ppi, 10 ppi, 20 ppi and 40 ppi, ranging from about 3×10^6 to 27×10^6 mesh elements. The numerical results are in good agreement with measurement data, with deviations of about 5 % to 15 % in particular.

In the present chapter we therefore employ the coupled phase field fluid flow and heat transfer method presented above, on real world foam

---

[3]http://www.ansys.com/ (accessed: 15/5/2014)

structures, whereas the experimental results provided in chap. 3 are used as validation data.

7.1 Numerics

As mentioned in chapter 5, the lattice Boltzmann method can be considered as being the discretised continuous Boltzmann equation, which asymptotically recovers the Navier Stokes equations. Since the method is formulated in a six dimensional phase-space, the spatial as well as the velocity space needs to be discretised. Hence, eq. (5.6) is discretised utilising either a two dimensional D2Q9 lattice model or a three dimensional D3Q19 model, cf. sec. 5.3. The discrete lattice Boltzmann equation (5.15) can be viewed as a finite difference representation of the continuous Boltzmann equation with simple explicit forward Euler time discretisation. Thereby, the lattice Boltzmann methods requires a regular Cartesian equally spaced grid.

For the heat transfer calculation equation (6.44) is solved in PACE3D on the same structured, equally spaced Cartesian grid, by means of an explicit finite difference formulation. At this point, it is crucial to note that the formalism of [168], which we have compared to regarding the sharp interface approach, is based on a steady state solution, whereas the accuracy and stability of a numerical time dependent solution will also depend on the choice of the discrete time step. In the course of model development we are using a rather simple fully explicit finite difference framework, employing fully explicit forward Euler time discretisation.

Whereas the first implementation of the tensorial approach given in sec. 6.5 is done by means of a MAC (marker and cell) scheme with a staggered grid arrangement with respect to the three normal components and three shear components of the symmetric tensors, the segmented approach outlined in sec. 6.7.3 is less involved. For the latter, a central discretisation scheme and an upwind scheme as well as a total-variation diminishing (TVD) scheme [101] are used for the diffusive and the pseudo-advective parts, respectively.

All the used discretisation schemes are standard schemes, which are well documented in the open literature [6, 73, 168, 200], hence a lengthy outline is omitted here. The grid resolution is chosen problem dependent, according to the rule of thumb that at least 6 to 10 cells should be used to represent the interfacial region. Timestep size is chosen with respect to the well known stability and quality criterions given in the literature [101, 102].

Above fluid flow and heat transfer solvers are coupled with the phase field solver of the software package PACE3D[4], developed at the Karlsruhe University of Applied Sciences, Institute of Materials and Processes and the Karlsruhe Institute of Technology, Institute of Applied Materials. It allows for the simulation, visualisation and analysis of problems subject to phase transitions or interface dynamics, based on the multicomponent multiphase phase field method of [154]. Besides the basic evolution equations of phase field, concentration and energy, the coupling of additional solvers for elasticity, magnetism of fluid flow allows the realisation of multidisciplinary simulations. However, in the course of this work, focus is given exclusively to the development of the fluid flow and heat transfer capabilities, whereas no phase transition or interface dynamics are considered herein. The software PACE3D and the methods outlined in the preceding chapters is parallelized using OpenMPI[5], whereas a one-dimensional and three-dimensional domain decomposition allows for extremely efficient performance on high performance computing (HPC) architectures, cf. [103, 211].

7.2 Foam modelling using **PACE3D**

Depending on the manufacturing process most cellular solids in general, and in particular open cell metal foams are intrinsically stochastic structures. Therefore, regular unit cell based structures most often lack from the true features of the microstructure and its effects.

In sec. 2.5 the algorithm developed by [178] is described, which is used in the following for the creation and modelling of open cell metal foam structures. The method is controlled using a statistical measure – mean and standard deviation – of the pore diameter and the edge diameter, whereas the porosity and specific gravity are results of the final structure. Due to its heuristic nature, the presented algorithm is able to reflect the stochastic nature of the foam structures and therefore represent the flow and heat transfer properties more realistic, cf. fig. 7.1.

---

[4]Parallel Algorithm for Crystal Evolution in 3D
[5]http://www.open-mpi.de (accessed: 10/3/2014)

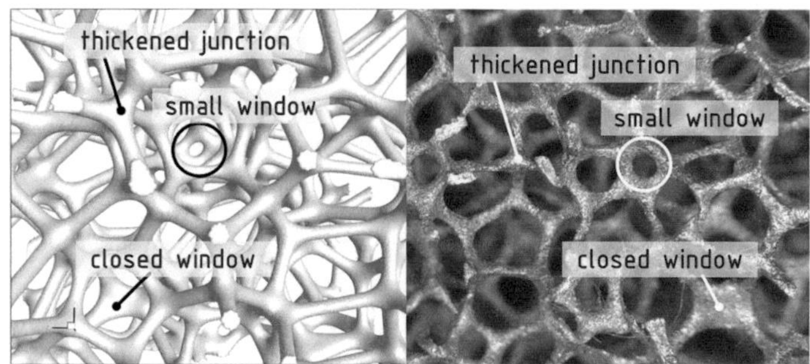

Figure 7.1: Pore scale features (small window, closed window and thickened junctions) of artificial and real foam structures.

The initialisation of the phase distribution is done by a sharp interface setup. At first we perform a phase-field simulation without driving forces, in order to establish a diffuse interface with finite width λ. The pre-calculations are stopped as the phase-field landscape becomes stationary.

The experimental data for foams of the same pore density showed significant differences in pore scale measures and porosity. For the as-

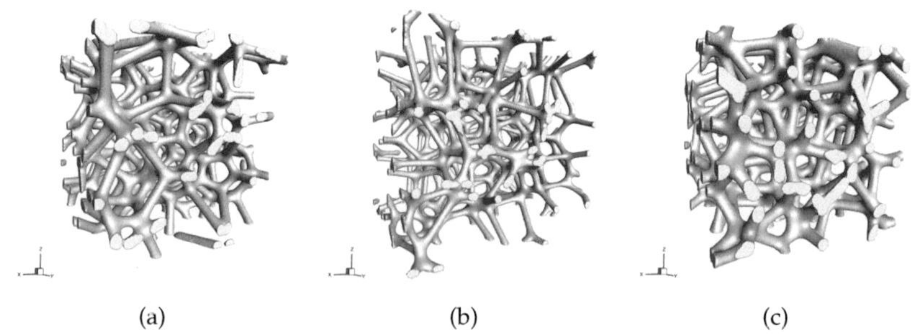

(a)   (b)   (c)

Figure 7.2: Different artificially generated foam structures with different porosities, to represent 10 ppi aluminium foams: (a) $\psi_{\text{modelling}} = 88.44\,\%$ corresponding to sample №1, (b) $\psi_{\text{modelling}} = 92.78\,\%$ corresponding to sample №2 and (c) $\psi_{\text{modelling}} = 83.85\,\%$ corresponding to sample №4, cf. tab. 7.1.

sessment of the numerically modelled foam structures emphasis was laid on capturing the porosity as realistic as possible. Since the basic packing of spheres in the foam modelling algorithm is based on a heuristic but reproducible model, the same set of input parameters can lead to foam models of different porosities. Thus, the seed-point of the heuristic model is varied over a certain range, and the suitable porosity, with respect to the experimental values, are found by an automated trial and error procedure.

Table 7.1: Comparison of the porosity of experimental and artificial foam structures.

sample	experiment	modelling	error
№ 1	88.441 %	88.440 %	−0.001 %
№ 2	92.777 %	92.780 %	+0.003 %
№ 3	87.707 %	87.700 %	−0.007 %
№ 4	83.846 %	83.850 %	+0.004 %
№ 5	87.889 %	87.890 %	+0.001 %

Table 7.1 exemplarily shows the comparison of the porosities for the real-world and the artificial foams for the 10 ppi aluminium samples used in the experimental investigations, cf. chap. 3, whereas three different artificial foam structures are depicted in fig. 7.2. The agreement in porosities given in tab. 7.1 is excellent, and the qualitative compliance of the detailed pore structure between model and microscopic images taken during the experiments is astonishing. Furthermore it must be mentioned here, that the process of foam generation is a matter of several seconds up to several minutes depending on the domain size. For a fine spatial resolution of 200^3 cells, the foam generation including a couple of phase-filed iterations for the development of a smooth interface, require about 6 min on an ordinary single CFD workstation.

7.3 Fluid flow and heat transfer simulation in cellular solids using StarCCM

For comparison in terms of quantitative results as well as for the qualitative assessment of the modelling procedure, a number of selected

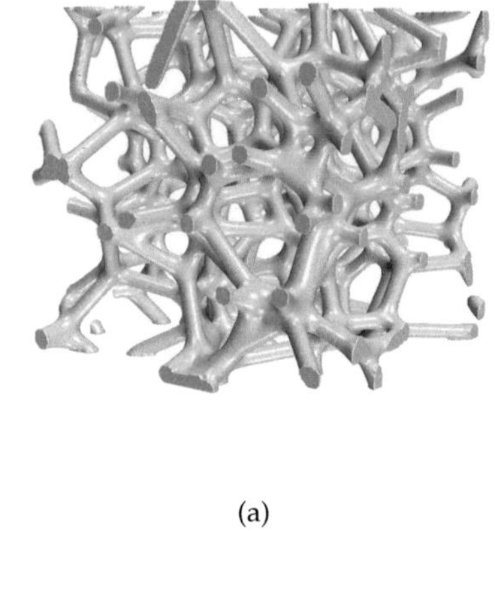

(a)

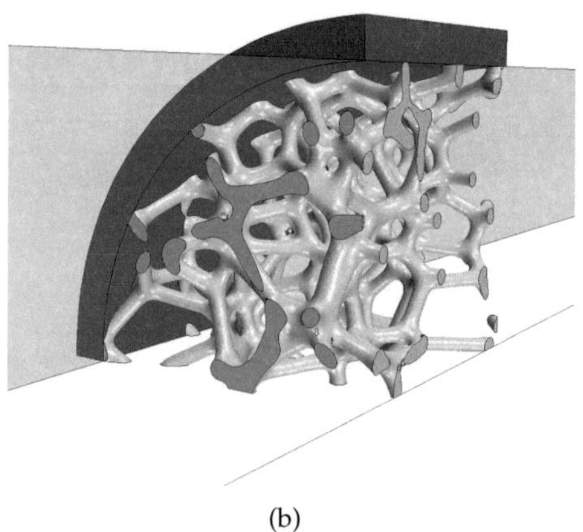

(b)

Figure 7.3: (a) Basic foam geometry (dimensions 2.2 cm × 2.2 cm × 2.2 cm) imported in StarCCM (b) customised sample geometry reflecting the real sample geometry and flow situation.

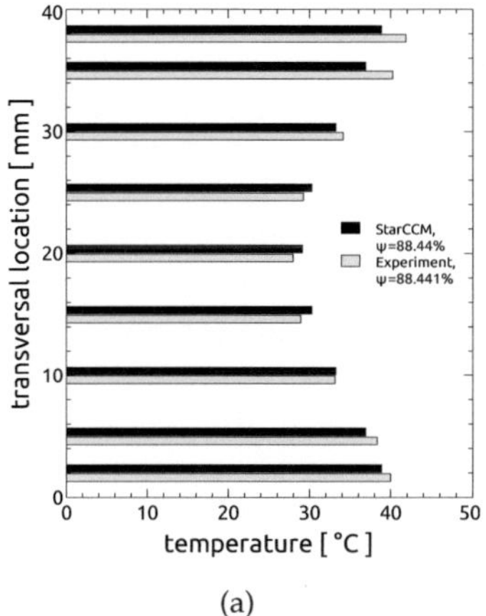

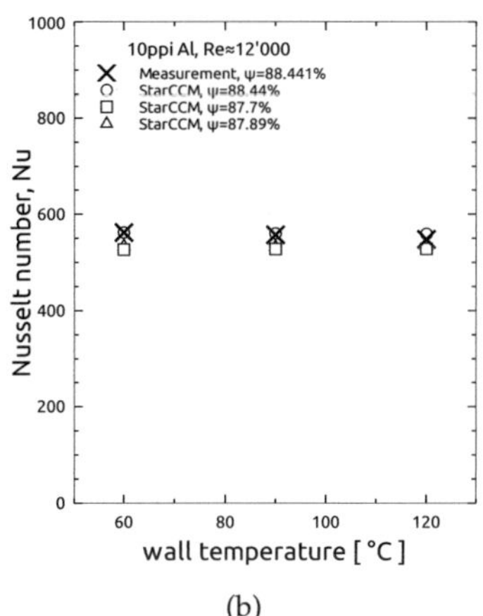

Figure 7.6: Exemplary thermal results of fluid flow and heat transfer simulation of open cell metal foam using StarCCM: (a) comparison of measured and simulated temperature profile in the outflow and (b) comparison of measured and simulated Nusselt numbers at three different temperature levels ($60°C$, $90°C$ and $120°C$) for foams of comparable porosity.

direction along the centerline. Thus, we start from a foam model with dimensions $2.2\,cm \times 2.2\,cm \times 2.2\,cm$, which needs to be further worked up, in order to reflect the real foam specimen, cf. fig.7.3. A suitable inlet and outlet length is modelled, to provide a fully developed flow in front of the sample and to avoid boundary effects on the flow leaving the foam. For conjugate heat transfer simulation, the heat source with respect to the experimental setup is modelled by a cylindrical solid shroud.

A high quality mesh is made from polyhedral cells, where prism layers are inserted in the vicinity of solid surfaces, resulting in 8.4×10^5 cells, 4.9×10^6 faces and 4.5×10^6 vertices. A longitudinal cross section is shown in fig. 7.4(a), whereas the solid surface of the foam and shroud is given in fig. 7.4(b). Appropriate boundary conditions are applied on all faces of the computational domain, and material properties of aluminium and air are used for fluid and solid. Finally, the Navier-Stokes and energy equations are solved in StarCCM by means of a coupled implicit finite volume method.

Exemplary results for the pressure loss is given in fig. 7.5 for a 10 ppi aluminium foam sample of 20 mm length. As indicated in sec. 7.2, the modelled foam structures are in very good agreement with the measured samples in terms of porosity, cf. tab. 7.1. The simulated and measured pressure losses are in good agreement, whereas slightly higher deviation is observed in case of high Reynolds numbers. The numerical results are within an constant error band of about $\pm\, 1.5\,\%$, except for the highest Reynolds number, where the difference is in the order of about 4%. Despite of the excellent agreement of the foams in terms of porosities, structural and topological differences between the real and artificial foams may account for the deviations in pressure losses.

Figure 7.6 shows exemplary results of the thermal simulations using the artificial 10 ppi aluminium foam structures. Even though the modelled foam structure does not coincide with the real foam structure in all pore scale details and features, the temperature profile of the flow downstream of the sample is in rather reasonable agreement with the measurement, cf. fig. 7.6(a). Furthermore, the integral heat transfer performance in terms of Nusselt number is given in fig. 7.6(b) for three different artificial foam samples with porosities $\psi = 88.44\,\%$, $87.7\,\%$ and $87.89\,\%$ compared to the measured values for the 10 ppi aluminium sample of porosity $\psi = 88.441\,\%$. Whereas the simulation shows excellent agreement for the simulation using an artificial foam of almost identical porosity ($\psi_{\mathrm{sim}} = 88.44\,\%$ compared to $\psi_{\mathrm{exp}} = 88.441\,\%$), the simulation for two foams of smaller porosities yield slightly decreased Nusselt numbers.

The overall agreement of experimental and numerical fluid flow and heat transfer features is promising, and demonstrates that the artificial foam structures represent a suitable geometrical modelling approach.

7.4 Fluid flow simulation in cellular solids using **PACE3D**

In the following we present results of the coupled lattice Boltzmann and segmented tensorial heat transfer solvers outlined in chapters 5 and 6, respectively. The foam structures mentioned in sec. 7.2 are used, amongst other foams which are generated with respect to the measured samples, cf. sec. 3.2. We will focus on a numerical domain that represents an open cell metal foam probe of dimensions $\varnothing\, 40\,\mathrm{mm} \times 20\,\mathrm{mm}$.

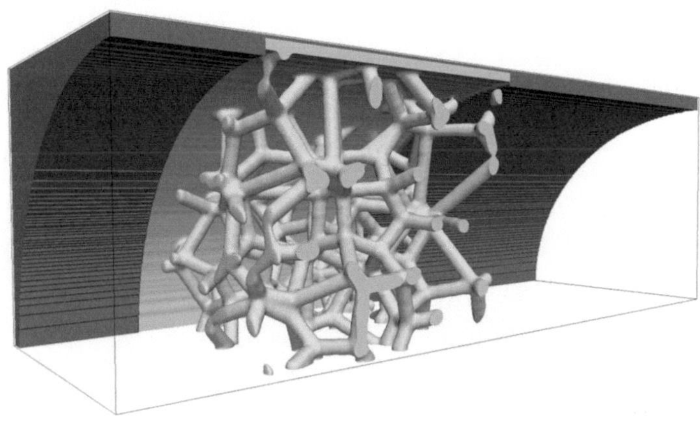

Figure 7.7: Computational domain of $200 \times 200 \times 600$ cells used in PACE3D. The foam is depicted as light greyish surface, whereas the dark grey region indicates the outer solid margin that is defined as a so called barrier region which does not account for the numerical computation.

In contrast to StarCCM the coupled solvers in PACE3D make use of the same uniform and equally spaced Cartesian grid. With respect to a suitable resolution of the finite interfacial region between fluid and solid phases – from experience about 6 to 10 cells – different numerical grids ranging from 100×100 to 200×200 cells in lateral and 400 to 600 cells in streamwise direction are generated. Thanks to the cylindrical shape and in order to save computational resources and simulation time, one quarter of the sample is modelled, expecting two symmetry planes along the flow direction. The geometrical representation of the foam is given in fig. 7.7 shaded in light grey, whereas the outer solid tubular margin is indicated as dark grey region. The latter is defined as a so called *barrier* region, that does not account for the computation and efficiently reduces the numerical effort. Thus, the overall amount of active cells for the numerical computation reduces from about 24×10^6 to about 18×10^6. A cylindrical inlet and outlet region is modelled, and it is proven that we get a developed velocity profile at the inlet, whereas extrapolation boundary conditions are used at the outlet of the numerical domain for fluid flow and heat transfer, respectively.

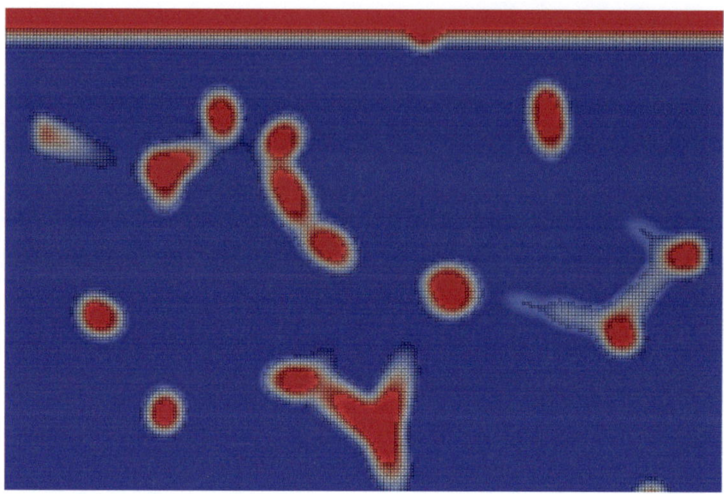

Figure 7.8: Distribution of the order parameter on a longitudinal cross section through the computational domain. The blue regions indicate fluid, the red regions solid and the diffuse interface is highlighted by showing the corresponding cells in case of non vanishing gradients of the order parameter.

Whereas fluid flow simulations are only carried out in the fluid region of the computational domain, the coupled heat transfer and fluid flow calculations are seamlessly applied on both, the fluid and solid regions. it should be pointed out, that by using the phase field method no explicit boundary or coupling conditions have to be defined at the fluid-solid interfaces. For the mentioned barrier region no calculation is executed, and the interface between barrier and an arbitrary phase acts like an adiabatic solid wall. Since the barrier region is not modelled as a diffuse interface, the tubular geometry results in a staircase representation (dark grey), whereas the cylindrical shroud of the foam is modelled as solid phase, which results in a smooth cylindrical shape. However, since drag and friction of the foam by far exceeds the drag and friction induced by the cylindrical inlet and outlet region we accept this shortcoming.

Starting from a sharp, staircase like representation of the structure, the diffuse interface is established by performing a couple of phase field

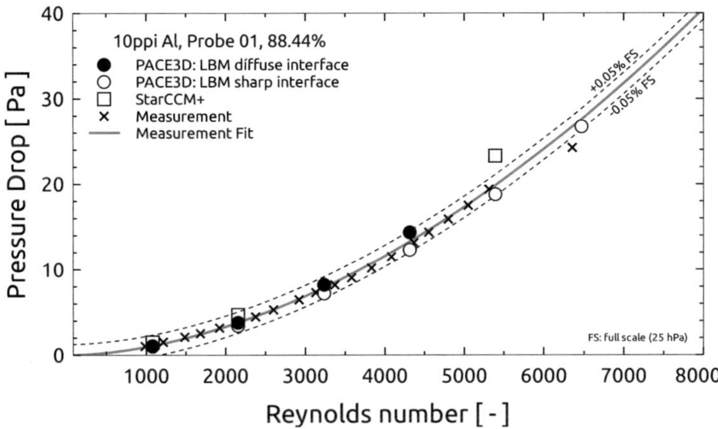

Figure 7.9: Comparison of measured and simulated pressure loss data. An additional measurement series is performed for low to medium Reynolds numbers using a high accuracy pressure transducer. The level of accuracy is indicated by the upper and lower error limits at $\pm 0.05\%$. The results of StarCCM are shown as a reference, whereas *sharp interface* as well as *diffuse interface* results are given for PACE3D.

iterations; without a driving force, to avoid growth of shrinkage of the structure. Figure 7.8 shows the distribution of the order parameter on a cross section of the computational domain. The blue regions indicate fluid, the red regions solid and the diffuse interface is highlighted by showing the corresponding cells in case of non vanishing gradients of the order parameter.

Figure 7.9 depicts the experimental and numerical integral pressure loss for a foam probe with a pore density of 10 ppi and a porosity of 88.44 % for different Reynolds numbers. In addition to the measurements outlined in chapter 3 an additional experimental series is done with respect to detailed pressure losses at low to medium Reynolds numbers ranging from about 800 to 6'500. These measurements are obtained using a high precision pressure transducer (Endress+Hauser) with an accuracy of $\pm 0.05\%$ for a full scale of 25 hPa, indicated by the upper and lower error limits (dashed lines) in the figure. The results of StarCCM are included for comparison, whereas two different simulation series

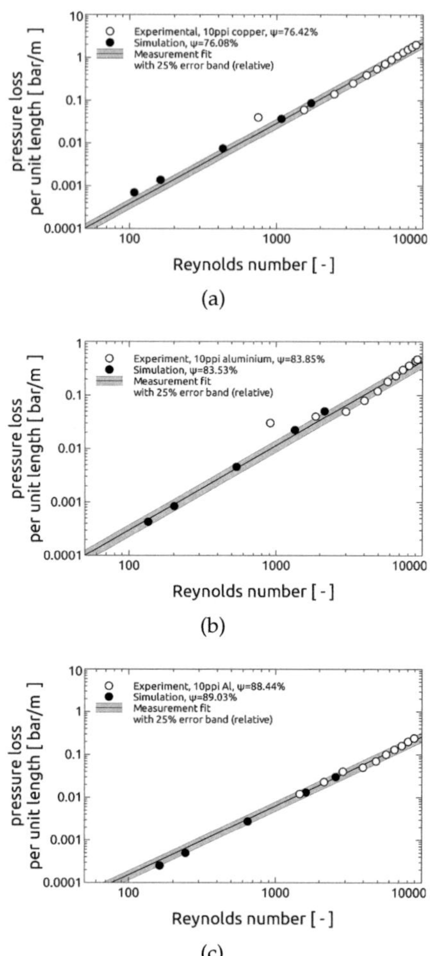

Figure 7.10: Pressure loss per unit length data over Reynolds number for different artificial foam structures with pore diameters of 4 mm, 5 mm and 6 mm and a constant edge diameter of 0.6 mm, which result in porosities of (a) $\psi = 76.08\,\%$, (b) $\psi = 83.53\,\%$ and (c) $\psi = 89.03\,\%$, respectively. The most equivalent foams in terms of porosity (10 ppi copper $\psi = 76.42\,\%$, 10 ppi aluminium $\psi = 83.53\,\%$ and 10 ppi aluminium $\psi = 88.44\,\%$) are chosen for comparison, whereas the fit of the experimental data is extrapolated towards low Reynolds numbers.

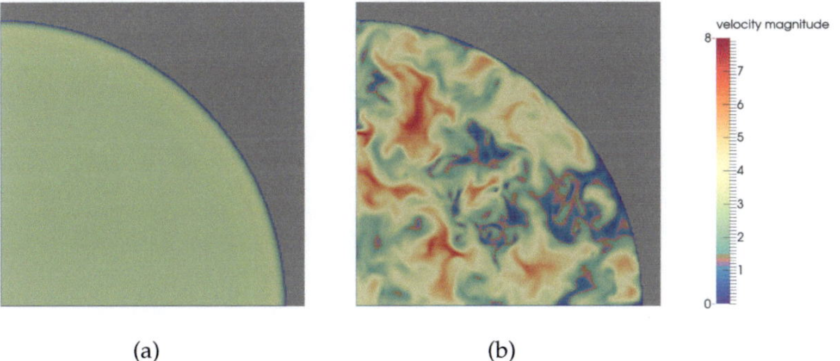

(a)   (b)

Figure 7.11: Distribution of velocity magnitude (a) up- and (b) downstream of the foam sample for a Reynolds number of about Re≈ 6'000.

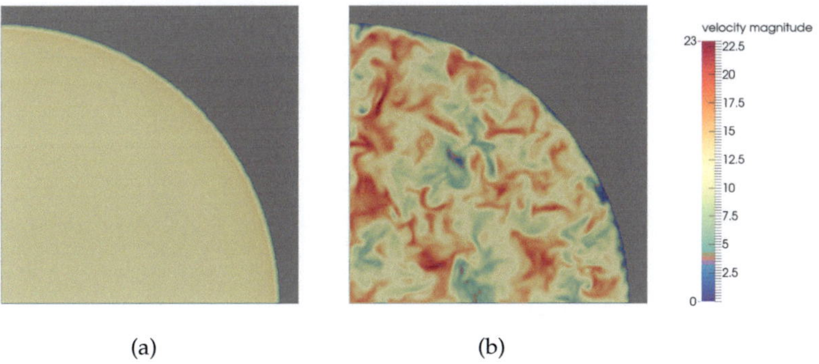

(a)   (b)

Figure 7.12: Distribution of velocity magnitude (a) up- and (b) downstream of the foam sample for a Reynolds number of about Re≈ 26'000.

are carried out using PACE3D. A *coarse* staircase mesh with dimensions $100 \times 100 \times 400$ cells is used for a *sharp interface* like solution, without applying smoothing with respect to the diffuse interface generation. On the other hand a finer mesh with dimensions $200 \times 200 \times 600$ cells is used for a *diffuse interface* simulation, where the interface spans about 6 to 10 cells. While the results of StarCCM show slightly higher deviations for higher Reynolds number, the results obtained with PACE3D, for both

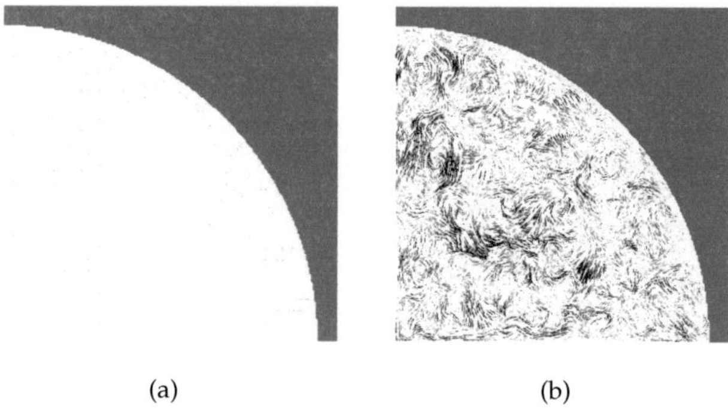

(a)　　　　　　　　　　(b)

Figure 7.13: In-plane vectors, depicting the secondary flow structure (a) up- and (b) downstream of the foam sample for a Reynolds number of about Re≈ 6'000.

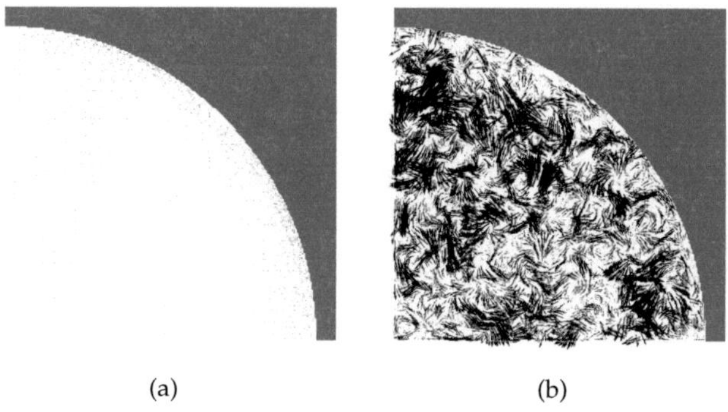

(a)　　　　　　　　　　(b)

Figure 7.14: In-plane vectors, depicting the secondary flow structure (a) up- and (b) downstream of the foam sample for a Reynolds number of about Re≈ 26'000.

sharp as well as *diffuse interface* setup, are all within the $\pm 0.05\%$ error margin.

Further numerical test are performed on artificial foam structures with pore diameters of 4 mm, 5 mm and 6 mm and a constant edge diameter of 0.6 mm, which result in porosities of $\psi = 76.08\%$, 83.53% and 89.03%, respectively. The results of flow simulations with Reynolds numbers ranging from about 150 to about 3000 are compared with the experimen-

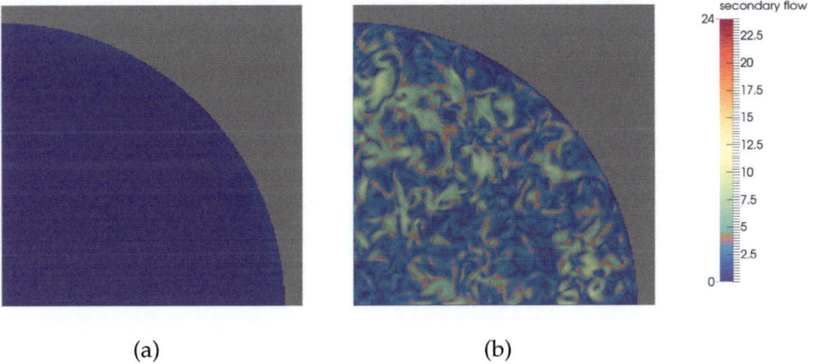

(a)   (b)

Figure 7.15: Velocity distribution of the secondary flow structure (a) up- and (b) downstream of the foam sample for a Reynolds number of about Re≈ 6'000.

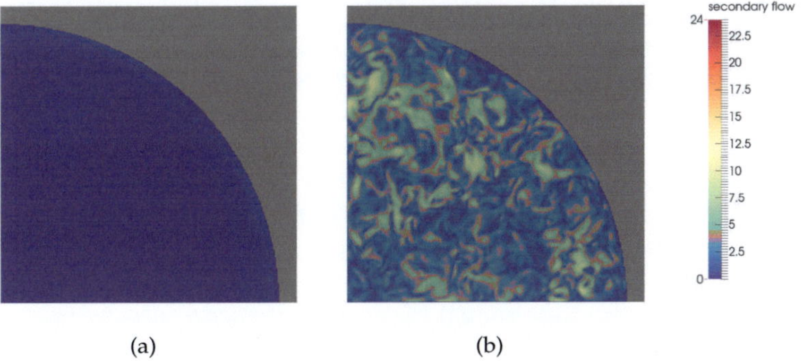

(a)   (b)

Figure 7.16: Velocity distribution of the secondary flow structure (a) up- and (b) downstream of the foam sample for a Reynolds number of about Re≈ 26'000.

tal data of the most comparable foam structures in terms of porosity used in the experimental investigation, i.e. 10 ppi copper $\psi = 76.42\,\%$, 10 ppi aluminium $\psi = 83.53\,\%$ and 10 ppi aluminium $\psi = 88.44\,\%$. The results are depicted in figs. 7.10 in comparison to the extrapolated measurement fits.

Apart from comparing integral performance values in terms of pressure loss, the simulation of complex cellular structures provides a high level of details, which can hardly be accessed by experimental methods, if ever. Thus, in the following a couple of flow features are visualised and discussed. The distribution of velocity magnitude is given in figs. 7.11 and 7.12 on a lateral cross section up- and downstream of the foam sample for a moderate Reynolds numbers of about 6'000 and 26'000, respectively. There is an obvious difference in the velocity gradient next to the cylindrical wall for the velocity distributions upstream of the foam sample, that correctly corresponds to the Reynolds number, i.e. the flow that pertains to the higher Reynolds number shows a higher gradient, therefore a much smaller boundary layer. Regarding the downstream velocity distribution, the flow with the smaller Reynolds number shows a vast region of low velocity, compared to more homogeneous flow structure which belongs to a higher Reynolds number. Besides increasing turbulent effects, with increasing Reynolds number, the mixing of the flow due to vortex shedding, detachment of wakes behind the foam edges and higher frequency of unsteady flow phenomena is more distinctive, thus leading to a more *granular* velocity distribution.

This is acknowledged by the secondary flow structure, depicted by the plane-parallel secondary flow vectors given in figs. 7.13 and 7.14, as well as the distribution of the magnitude of secondary flow given in figs. 7.15 and 7.16. Despite of some inflow effects, due to the constant velocity boundary condition at the inlet, the secondary flow is vanishing upstream of the foam sample, cf. 7.13(a) and 7.14(a). On the other hand, the secondary flow structure downstream of the foam is likely to coincide with the topological features (pores and pore windows) of the foam. However, the flow structure belonging to a higher Reynolds number not only shows higher magnitude in secondary flow, but also the flow is more dispersed, cf. 7.15(b) and 7.16(b).

The occurrence of unsteady flow phenomena like vortex shedding in the wake of the foam edges can be identified in figs. 7.17(a) and 7.17(b). The flow with Re $\approx$ 6'000 shows large wakes and detached flow regions. The regions of separated flow downstream of the foam is more pronounced, whereas for the higher Reynolds number, the separation is more dispersed. The complex interaction of the different effects leads to a heavily distorted, fluctuating flow structure in the downstream region, cf. figs. 7.17(a) and 7.17(b). The longitudinal velocity distribution of the flow with higher Reynolds number shows unsteady flow features (vortex shedding) already downstream of the first foam edges. Besides

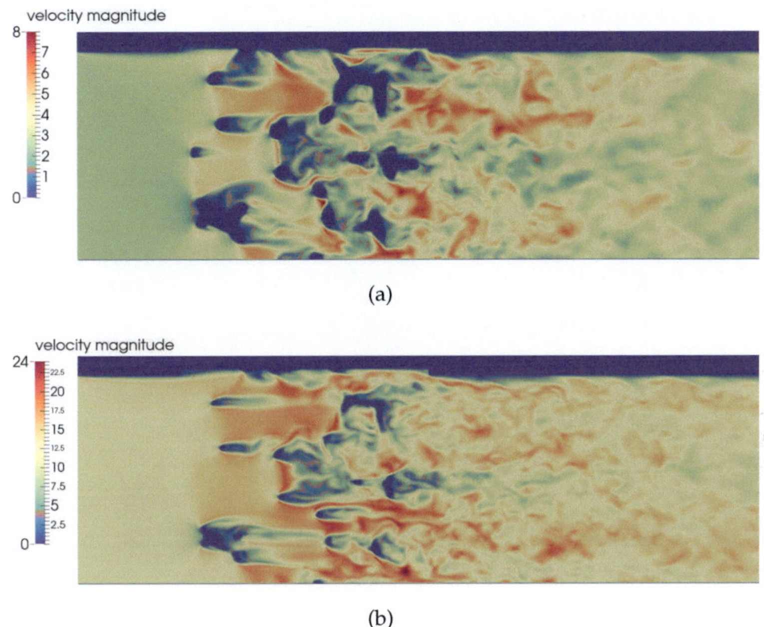

Figure 7.17: Velocity distribution on a longitudinal cross section for (a) a Reynolds number of about 6'000 and (b) 26'000 for a 10 ppi aluminium foam sample with porosity $\psi = 88.44\%$.

the frictional losses due to the formation of boundary layers, the main source of irreversible dissipation of energy is due to the eddies formed by the wake flow. Since the applied lattice Boltzmann method solves for the transient flow field, the unsteadiness will not disappear or smeared as for a steady RANS[8]-solver. However, for both Reynolds numbers under investigation, the fluctuating flow is mixed out not far downstream the foam – a well homogenised flow is identified just about two characteristic diameters downstream of the samples. Due to the unsteadiness of the flow, the analysis of the integral pressure loss as well as the thermal performance values in the subsequent section, are conducted as time averaged quantities.

---

[8]Reynolds Averaged Navier Stokes

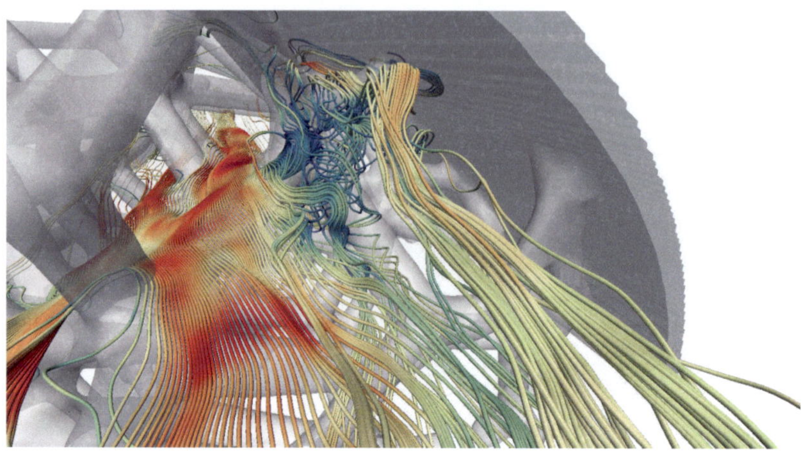

Figure 7.18: Visualisation of the flow inside the foam structure with three-dimensional streamlines at a Reynolds number of about 26'000. Besides straight flow paths, detached flow is identified in the wake of a foam edge. Streamlines are coloured by the magnitude of the velocity.

Finally, a three dimensional visualisation of streamlines is given as an example for the level of details which is provided by a numerical simulation at pore scale level. Figure 7.18 shows some exemplary streamlines inside the foam structure, coloured with the local velocity magnitude. Whereas some of the streamlines pass through the structure on a straight path and without significant disturbance, some streamlines close to the wake of a foam edge give some idea of the separation and recirculation that takes place. Adjacent streamlines of high and low velocity magnitudes will form free shear layers; again a source of dissipative energy loss.

7.5 Thermal simulation of cellular solids using **PACE3D**

In this section we provide results for the simulation of forced convection inside a foam structure. The flow is driven externally, and dominates buoyancy effects. According to the ratio of Grashof number and square of the Reynolds number three regimes are classified, where different effects will dominate the coupled flow and heat transfer, cf. tab. 7.2.

Table 7.2: Classification of coupled fluid flow and heat transfer according to the ratio of Grashof and square of Reynolds number.

condition	regime
$Gr/Re^2 \ll 1$	domination of forced convection
$Gr/Re^2 \sim 1$	same order of magnitude of buoyant and inertial forces
$Gr/Re^2 \gg 1$	domination of natural convection

With a ratio of $Gr/Re^2 < 1$ for almost all of our measurements and $Gr/Re^2 < 0.1$ for the selected application, buoyant effects can be neglected. From the fact that we operate in a moderate regime close to ambient conditions regarding temperature and pressure, and from the experimental results in chap. 3, we know that the Prandtl number is at about 0.7. As a consequence, the thermal and momentum boundary layers are of the same order, i.e. no additional refinement of the mesh is required. However, the application of the segmented tensorial approach on such complex geometry is a challenging task.

As a first qualitative test application, we consider pure heat conduction in an open cell metal foam. The foam structure is a computer generated geometry model mounted on a base plate, which is subject to a constant temperature at the bottom. The physical properties of aluminium are applied on the foam, while the surrounding is modelled using the physical properties of air. Note, that forced as well as free convection is omitted here. This case represents a rather small foam structure including only a few struts and pores. We consider a uniform rectangular domain of $0.02\,\text{m} \times 0.02\,\text{m} \times 0.02\,\text{m}$, which is discretised with a rather coarse uniform rectangular grid of $80 \times 80 \times 80$ cells. A temperature difference of $\Delta T = 60\,\text{K}$ is applied on the base plate, compared to the remaining domain which is initialised at $T = 300\,\text{K}$. The simulation is carried out for a physical time of $t = 5\,\text{s}$, and fig. 7.19 shows qualitative visualisations of the results of the computations of the foam structure, where the colour bar indicates the normalised temperature. The surface temperature of the ligaments is given in 7.19(a), and fig. 7.19(b) depicts the temperature distribution in the surrounding air on an imprinted transparent cube.

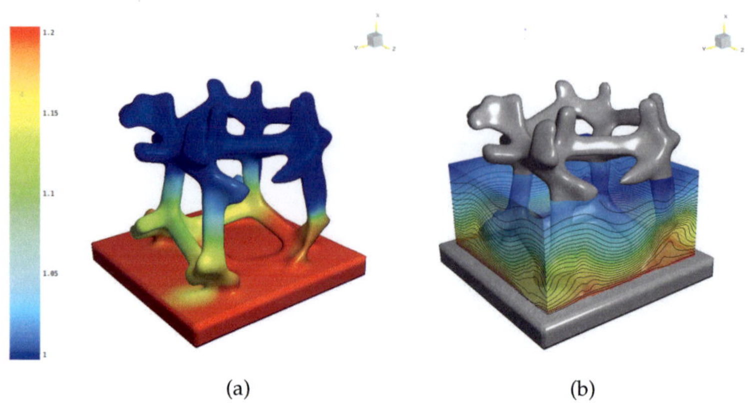

Figure 7.19: Temperature distributions on a wall-mounted open cell metal foam with constant wall temperature: (a) temperature distribution on the ligaments of the aluminium foam (b) temperature distribution and isolines on an imprinted transparent cube.

In absence of experimental and analytical reference data, a qualitative examination yields, that the results are reasonable and that the segmented tensorial approach is suitable for the thermal simulation in complex convoluted three dimensional cellular structures. Especially from the numerical point of view, the approach is found to be robust and stable, no stability issues, spurious effects or oscillations are observed.

A hybrid thermal model (HTLBE-Model[9]), where a lattice Boltzmann fluid flow method is coupled with a classical finite difference heat transfer algorithm, is already proposed by [120], and successfully applied by [108, 153, 206], to mention only a few. By decoupling of mass and momentum from energy conservation [207], it is assumed that for a weak coupling of fluid flow with temperature allows for using an athermal lattice Boltzmann model for the solution of the flow field, whereas the temperature field coupled by convective and buoyancy terms [146]. With respect to [206], the acoustic modes in the temperature equations that refer to real gases are neglected in the course of the present work. Thus, we solve for the incompressible energy equation and no scaling of the

---

[9]**Hybrid Thermal Lattice Boltzmann Equation Model**

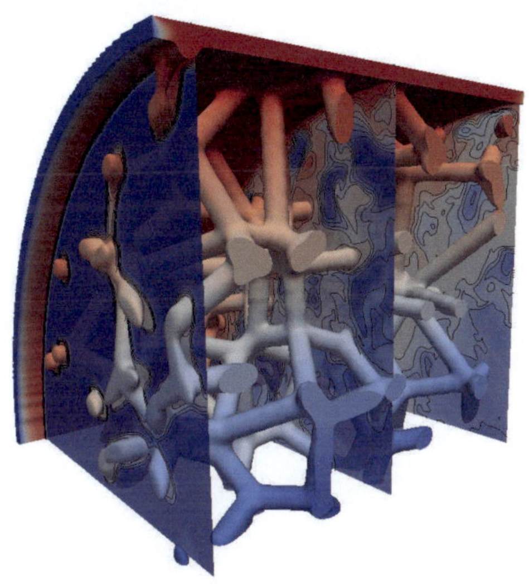

Figure 7.20: Temperature distribution inside a 10 ppi aluminium foam sample of porosity $\psi = 88.44\,\%$. The temperature is given on the surface of the foam structure as well as on three lateral cross sections for the fluid, where low and high temperature corresponds to blue and red colours, respectively.

thermal diffusivity is required [120]. Due to the different stability criteria of an advection diffusion process, namely the CFL condition[10] and grid Peclet number[11] [101], and for the lattice Boltzmann method (eq. (5.21)), the time step size is evaluated accordingly and, if necessary, chosen differently for the fluid flow and heat transfer part.

To finally test the coupled implementation of the diffuse interface fluid flow and diffuse interface heat transfer models on a real world engineering type application, a 10 ppi aluminium foam sample of 20 mm length is simulated. The computational domain is equal to the one used for fluid flow simulation, cf. fig. 7.7. In doing so, the solid phase of

---

[10]Courant Friedrichs Lewy condition: for explicit time integration schemes, the time step width should be less than the time for the information to travel across the computational grid.

[11]For a stable explicit time integration scheme, it is sufficent (not necessary) that the grid Peclet number Pe= $u\Delta x/\alpha < 2$.

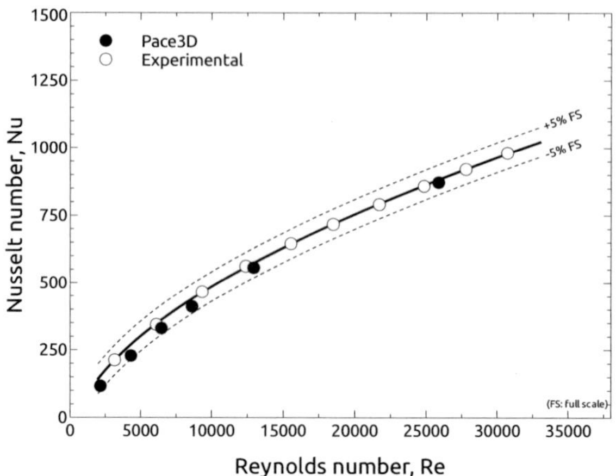

Figure 7.21: Comparison of numerical and experimental Nusselt numbers for a 10 ppi aluminium foam with a porosity of $\psi = 88.44\,\%$.

the foam is embedded into a cylindrical barrier tube which is used to model the adiabatic solid walls of the circular cross section flow channel. Thanks to the phase field approach, no extra modelling or definition of the fluid-solid interfaces is necessary. A special so called *barrier condition* is defined for the spatial region of the cylindrical shroud of the foam, which applies a constant wall temperature to a certain phase which is adjacent to the barrier, according to the experimental setting. Figure 7.20 exemplarily shows the heated foam structure with three cross sectional planes depicting the temperature distribution of the fluid. On the outer cylindrical shroud (coloured red according to high temperature), the boundary condition applies, and is almost constant all over the shroud geometry due to the high thermal conductivity of the bulk solid material, whereas the foam structure shows a smooth radial temperature gradient. Compared to this, the temperature gradient in axial direction inside the foam is hard to see, again, due to the high thermal conductivity compared to the surrounding fluid.

We conduct a series of simulations in order to compare the integral thermal performance by means of the Nusselt number for different Reynolds numbers. Figure 7.21 shows the results compared to the experimental

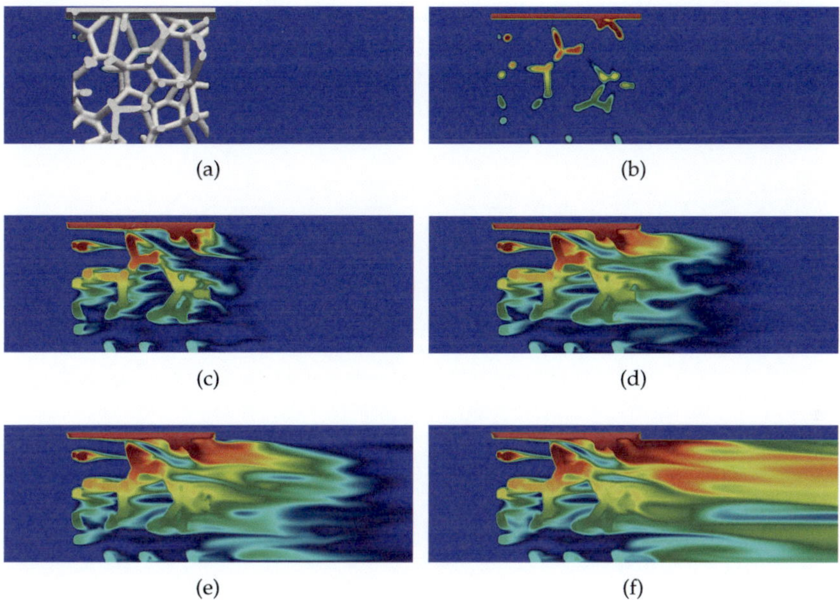

Figure 7.22: Time series of unsteady temperature distributions on a longitudinal cross section. Here, the foam structure depicted in (a) is subject to a radial initial temperature distribution (b). The figs. (c) to (f) give an impression of the temporal evolution of the temperature field.

reference values gained from the studies in chap. 3. According to the measurements, where the recorded physical quantities are time-averaged, the simulation results are averaged over a representative period. The results of the simulation are all within an error band $\pm 5\%$ of the full scale value. The overall agreement of the numerical and experimental reference values is excellent for the contemplated range of Reynolds numbers.

Finally, to give an idea of the temporal evolution of the temperature field, we provide a small series of consecutive snapshots. Figures 7.22 provide a view on the foam structure and the initial solution of the temperature field in figs. 7.22(a) and 7.22(b), respectively. These are followed by four timesteps showing the temperature field on a longitudinal cross

section of the computational domain. The unsteady flow phenomena discussed above are also obvious in the pictures of the evolving temperature
field in figs. 7.22(c) to 7.22(f), respectively.

Chapter 8

Outlook

Within this chapter we provide a sneak preview on some topics that emerged during the work on this thesis and were found to be worth focusing on in future developments and optimisations of the physical modelling and the numerical methods.

8.1 Multiphase diffuse heat transfer

So far the formulation of the segmented tensorial approach given in eq. (6.60) sec. 6.7.3, is valid for a binary system, which is sufficient and feasible for the present work. However, it is aspiring, and in the context of a multicomponent multiphase phase field method it appears necessary, to extend or generalise the formulation for an arbitrary number of phases.

Thinking of a domain which is made from a number of $n > 2$ phases, we have binary interfaces as well as triple-points, quadruple-point and any further combinations, depending on the number of adjacent phases. The key issue is exactly the treatment of these connections points with $n > 2$ phases. As a first attempt we start from a ternary system with a resulting heat flux vector that prevails at the triple point. We further assume, that the resulting heat flux vector is composed from the individual combinations of the binary $\alpha|\beta$ sub-interfaces, which reads

$$j = \sum_{\alpha < \beta} j_{\alpha\beta} \, . \tag{8.1}$$

The corresponding flux is formulated similar to what was derived in chap. 6 as

$$j_{\alpha\beta} = \frac{\phi_\alpha \phi_\beta}{\sum_{\alpha < \beta} \phi_\alpha \phi_\beta} \left(k_\perp^{\alpha\beta} \mathbf{Q}_{\alpha\beta} + k_\parallel^{\alpha\beta} (1 - \mathbf{Q}_{\alpha\beta}) \right) \nabla T \, , \tag{8.2}$$

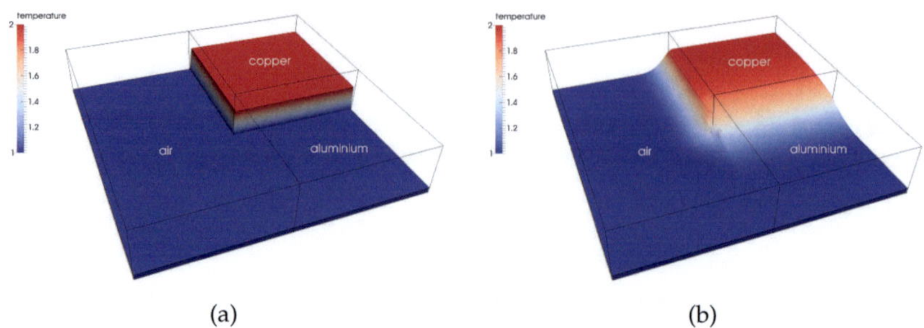

(a)   (b)

Figure 8.1: Exemplary application of the segmented tensorial mobility multi-phase formulation. Heat conduction in a ternary 2D domain: (a) initial temperature distribution and (b) snapshot of the steady temperature distribution.

where the leftmost fraction is a weighting factor which accounts for the volume fractions of the involved phases. The corresponding thermal conductivities $k_\perp^{\alpha\beta}$ and $k_\parallel^{\alpha\beta}$ are derived by the same interpolation schemes outlined in sec. 6.4. For a system of $n > 2$ phases, the projection $Q_{\alpha\beta}$ is given as

$$\mathbf{Q}_{\alpha\beta} = \frac{\boldsymbol{q}_{\alpha\beta}}{|\boldsymbol{q}_{\alpha\beta}|} \otimes \frac{\boldsymbol{q}_{\alpha\beta}}{|\boldsymbol{q}_{\alpha\beta}|} , \tag{8.3}$$

where the generalised interface normal reads

$$\boldsymbol{q}_{\alpha\beta} = \phi_\beta \nabla \phi_\alpha - \phi_\alpha \nabla \phi_\beta . \tag{8.4}$$

For the solution of the segmented tensorial formulation in eq. (6.60), the piecewise binary formulation and weighting is also applied on the formulation of the spatially varying divergence operator $\widetilde{\widetilde{\nabla}}_{\mathcal{R}}$.

A first test scenario is build from a three phase setup of copper, aluminium and air, depicted in fig. 8.1(a). Initially, the copper phase is at dimensionless temperature $T_{\mathrm{Cu},0} = 2$, whereas the aluminium and air phases are at dimensionless temperature $T_{\mathrm{Air},0} = T_{\mathrm{Al},0} = 1$, whereas all domain boundaries are set to an isolated condition ($\partial T / \partial n = 0$, with normal direction n). At time t_0 the phases are brought into contact, and the temperature evolution is monitored for all times $t > t_0$ until steady state is reached.

Whereas the first result looks quite promising, a closer look reveals that the temperature distribution is not smooth in the region of the triple-point, cf. fig. 8.2. Obviously, further development is necessary for the correct qualitative and quantitative multiphase treatment. A second attempt is to use the gradient of the mobilities, ∇M instead of the gradient of the order parameter $\nabla \phi$ in the definition of the projection (eq. (8.3)) and a different weighting.

Furthermore, the current formulation and implementation of the phase dependent decaying factorisation presented in sec. 6.4.6, causes a jump at the transition from bulk to interface. Even though this does not cause any problem until here, a more suitable formulation is preferable. We already conducted several tests using a sinusoidal formulation for the interfacial heat capacity $C_V(\phi)$, which reads as

$$
C_V(\phi) = \left[\frac{h(\phi_\alpha) + \phi_\alpha \phi_\beta \cdot f \cdot \sin\left(\frac{\pi}{8}\phi^m\right)}{C_V^\alpha} + \right.
$$

$$
\left. + \frac{h(\phi_\beta) + \phi_\beta \phi_\beta \cdot f \cdot \sin\left(\frac{\pi}{8}\phi_\beta^m\right)}{C_V^\beta} \right]^{-1} . \quad (8.5)
$$

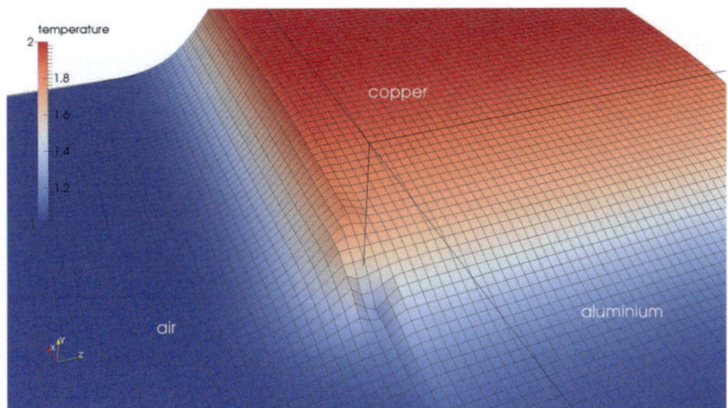

Figure 8.2: Close up of the temperature at the triple point, revealing a slightly crinkled distribution.

Here, $h(\phi)$ is the interface interpolation function and the factors g and m are subject for adaptation and optimisation of the interpolation function with respect to the ratio of C_V^α / C_V^β. This kind of formulation ensures the same integral factorisation, and at the same time provides a smooth, continuous and differentiable representation, which matches the bulk limits, cf. fig. 8.3.

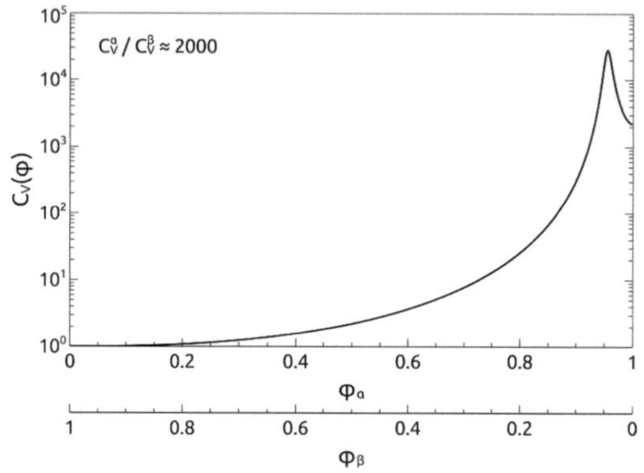

Figure 8.3: Potential candidate of interpolation function (eq. (8.5)) which guarantees a smooth, continuous and differentiable representation of $C_V(\phi)$ inside the interface.

However, these issues are provided here by means of an outlook, but they are subject of ongoing research and development and will be presented elsewhere.

8.2 Kinetic representation of heat conduction equation

A comprehensive overview on the properties of thermal lattice Boltzmann models is given in [119, 120], and the hybrid thermal lattice Boltzmann equation model (HTLBE-model) is established therein. The latter approach is applied by [146, 206, 207] and is successfully employed in the present work.

However, numerous authors report on the successful application of the so called passive scalar approach [137, 184, 195] and the double distribution function approach [1, 109, 147, 152, 201, 212], to name only a few. Apart from the disadvantage of poorer efficiency because of too many redundant degrees of freedom [119], it has been found that the method allows for larger timesteps [176]. The proposed approach is successfully applied within the context of phase field methods simulating dendritic growth and scalar diffusion. Due to the fact, that diffusion is modelled by a relaxation towards equilibrium rather than a second order spatial derivative, the timestep scales linearly with the grid size rather than quadratically [175].

8.3 Adaptive mesh refinement

The computations mentioned in chap. 7 are performed on equally spaced Cartesian grids, where the same spatial resolution is used for the complex convoluted microstructure of the foam and the up- and downstream regions. Since we do not expect significant gradients of the dominating variables (temperature, velocity, pressure) in the upstream region of the foam, the number of elements used for the computation could be dramatically reduced. Also for the microstructure or the downstream region, the spatial resolution could be controlled by the gradient of the flow or thermal variables. The mentioned idea is called *adaptive mesh refinement* (AMR) and is employed by many academic and commercial solvers in different disciplines. Although the performance (computational time per timestep and cell) remains constant, the computational costs are reduced, due to the scale dependent resolution of a simulation.

A successful and promising test implementation within PACE3D is provided by [116], utilising hierarchical grids, cf. fig. 8.4. At the interfaces of adjacent cells of different refinement levels, so called *hanging nodes*, interpolation and extrapolation techniques are necessary for a correct and feasible treatment of the physical quantities. Within the context of lattice Boltzmann methods, the interpolation is done between phase-spaces, which makes the numerical treatment and physical interpretation a bit more involved. However, different successful implementations are reported in the open literature. Whereas [39, 177] employ octree data structures for an efficient book keeping over the different grid levels, the methods given in [77, 203, 221] make use of hierarchical grids, which are more appropriate with respect to an implementation in PACE3D. For the

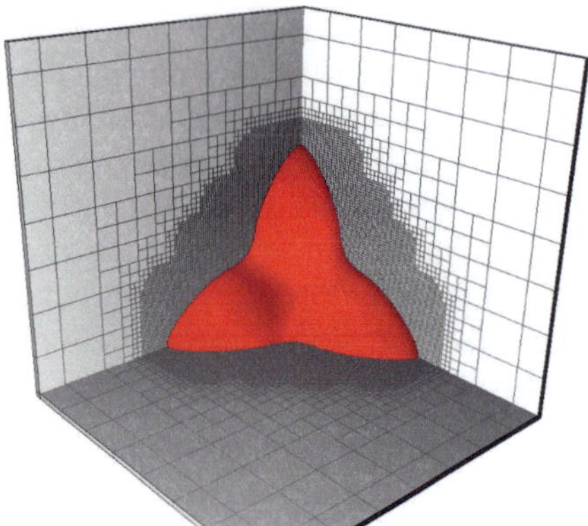

Figure 8.4: Exemplary visualisation of a simulation on crystal growth, employing a basic implementation of an adaptive mesh refinement method in PACE3D, picture taken from [116].

future it is desirable to extend the basic AMR implementation of [116] for the segmented tensorial mobility approach as well as for the lattice Boltzmann flow solver in PACE3D.

8.4 Multigrid convergence acceleration

In the LBM context, since the particle distributions move to the next lattice in one timestep, the well known CFL-number[1] is always equal to 1. As a consequence, convergence is mainly controlled by acoustic modes, which results in bad convergence for steady state problems [128]. When talking about adaptive mesh refinement above, the idea suggests to also employ a well established convergence acceleration technique, known as multigrid methods [73, 101]. By utilising several grids of different spatial resolution, high frequency errors are damped, which improves stability and convergence rate. There are numerous successful implementations reported in the open literature [101, 173], whereas [138, 204] report about successful implementations for lattice Boltzmann flow solvers.

---

[1]Courant-Friedrichs-Lewy number: stability condition: for explicit time integration schemes, the time step width should be less than the time for the information to travel across the computational grid.

Chapter 9

Conclusion

This work presents several contributions on the experimental and numerical investigation of fluid flow and heat transfer in cellular solids. In particular we focus on open cell metal foam, whereas the methods are generally applicable to all types of cellular solids. The results of the experimental work claims to enlarge the available hydraulic and thermal performance data of open cell metal foams, and at the same time it serves as validation data for the development of new and advanced numerical methods.

Rather than applying one of the classical academical or commercial numerical finite volume (FV), finite difference (FD) or finite element (FE) interface tracking methods, we based our models on a interface capturing phase field method [154]. A coupled diffuse interface fluid flow and heat transfer approach is established in view of dealing with even more complex geometries, and with regards to the dynamics of interfaces and convoluted multiphysics applications. One can easily think of topologies and configurations which can hardly be mastered by body fitted, interface tracking methods, if ever. For the latter situations, the presented method offers a feasible and comfortable way of performing detailed investigations.

Despite of the fact, that we have concentrated on problems related to fluid flow and heat transfer, the methodology is generally applicable to a wider range of applications such as mass transport, growth processes or coupling with solvers for elasticity or magnetism. In avoidance of unnecessarily convoluted examples, we focused on two phase systems with stationary steady interfaces. With this in mind, the numerical work presented herein claims to contribute to the continuous model development in the active and broad-ranging field of phase field methods.

With a focus on high performance and exa-scale computing, we were able to implement the mentioned numerical methods within the highly parallelized framework of PACE3D, employing a modern 3D domain decomposition approach with load balancing [211].

9.1 Experiments

We successfully performed hydraulic and thermal measurements for open cell metal foam samples of different pore densities, up to a maximum Reynolds number of about $\mathrm{Re} \approx 35'000$. Besides of acquiring validation data for the development of the numerical methods, the results gained from the experimental efforts contribute to enlarge todays database of pressure loss and heat transfer characteristics of cellular solid, by means of providing pressure loss data and Nusselt numbers up to a high Reynolds number flow regime. The accuracy, consistency and reliability of the measurements is assessed by multiple measurements and plausibility checks. Hydraulic and thermal performance is assessed by means of integral Hagen, Nusselt and Reynolds numbers.

Aluminium and copper specimens of different pore densities encased in cylindrical pipe segments were jointly manufactured from the same base material, respectively. A vast number of measurement series are carried out using customised measurement equipment and data acquisition. The modular configuration of the test rig as well as samples of equal pore density, facilitated the investigation of the hydraulic and thermal performance characteristics for the samples of diameter 40 mm and lengths 20 mm, 40 mm, 60 mm, 80 mm and 100 mm in streamwise direction.

Pressure loss characteristics are in qualitative agreement with comparable investigation of recent publications [53, 82, 132–134, 166, 222]. According to [47] a common pressure loss correlation is derived for all foam types, where the pressure loss is represented by the dimensionless Hagen number depending on the Reynolds number. With respect to the thermal measurements, we derived a Nusselt number correlation for each individual foam type as well as a common formulation for all foam types, based on well established formulations given in the open literature [18, 26, 40, 47, 54, 55, 75, 179, 218]. Comparison against simple tubular reference samples and the comparison of the fluiddynamic and thermal performance in terms of the hydraulic power and thermal resistance according to [17], gives an impression of the advantages gained from using open cell metal foams in heat exchangers.

9.2 Cellular Solids

The characterisation of the specimens in terms of statistical assessment and analysis of pore scale measures revealed the incapability of the commonly used pore density for the classification of open cell metal foams. Samples of equal pore density showed significant differences in pore scale measures and porosity, which is affirmed by the respective differences in hydraulic and thermal performance, cf. chap. 3.

In sec. 2.5 we presented an algorithm [178], which enables us to artificially generate realistic geometries of cellular solids. The astonishing realistic representation of foam structures is presented in chap. 7, where we were able to generate computational foam structures that match the real porosity except for deviations of less than a fraction of one percentage.

The complex microstructure geometries are generated automatically in minimal turnaround times. Using conventional CFD-packages, either structured or unstructured body fitted meshes are employed. In order to resolve complex microstructures, such as the open cell metal foams, these meshes require obvious user interaction and adaptation. Here, the representation of the microstructure in terms of a diffuse interface approach, provides a seamless and straightforward numerical apporach with minimum user interaction and minimal turnaround times.

9.3 Modelling of fluid flow

In chap. 5 we were able to present the successful coupling of single and multiple relaxation time lattice Boltzmann models with a phase field method. The linkage and the treatment of the diffuse interface region of the flow field is realized by two different models, the *reflectivity* [148] and the *forcing* model [10, 50, 157]. To avoid resolving the small scale turbulent effects by means of spatial and temporal discretisation, but to incorporate the effects for the application on high Reynolds number flow, the Smagorinsky subgrid scale turbulence model is implemented [33, 104, 121, 182, 187].

The results of the validation cases for plain poiseuille flow, flow in a lid driven cavity and for the turbulent high Reynolds number flow over a square cylinder at $Re = 20'000$ are in very good agreement with analytical and numerical reference solutions. Furthermore, it is obvious from these results, that the *forcing* model is superior to the *reflectivity* model.

9.4 Modelling of heat transfer

A comprehensive numerical survey on the transient two-phase heat conduction problem in the context of a diffuse interface method is given in chap. 6. Based on the motivating work of [158], where it is shown how the tensorial mobility approach eliminates thin interface effects, we extended this strategy for transient three dimensional two-phase heat conduction problems. The numerical survey of different interfacial interpolation techniques for the volumetric heat capacity has shown, that a factorised phase dependent interpolation gives best results. While the thermal conductivity is modelled as a tensorial quantity, a modified divergence operator is applied with respect to the different interpolation schemes used for different directions according to the interface.

Focusing on complex convoluted microstructure geometries, the detailed survey with regards to robustness and stability of the approach finally resulted in the segmented tensorial formulation. The mathematical derivation allows us to distinguish between quadratic and linear terms. Owing to the physical and pseudo-physical nature of the different terms, we apply dedicated discretisation schemes, whereby the final formulation results in a much simpler implementation compared to the staggered approach of the earlier model. Furthermore, all stability issues of the earlier formulation, related to high ratios in mobility and capacity properties, as well as complex high curvature geometries, are improved significantly.

We obtain excellent results for a number of basic one-, two- and three-dimensional validation cases, compared to analytical as well as numerical reference solutions, whereas different material combinations with low to high ratios in thermal conductivity and specific heat are tested. Note, that in the present work we performed quantitative simulations with

λ / R^1, which is ten times larger than given in [158]. Moreover, the ratios of material properties are also several orders higher. To the authors best knowledge, this is the first time [70] reporting on a transient tensorial modelling of diffusive transport problem within the context of phase field methods.

Whereas we utilised the phase-field method for the modelling of transient heat conduction for composite materials, the method is generally applicable to a variety of physical phenomena which obey classical diffusive transport equations. Moreover, the presented approach, implemented in PACE3D, is vast promising for transient free boundary multi physics problems, which are the scope of modern phase-field methods. Thus, the present model provides a good foundation for a variety of avenues for numerical modelling and application on multi physics problems, and the ongoing efforts should spread across different areas of interest.

9.5 Applications

Besides the validations carried out for the diffuse interface fluid flow approach outlined in chap. 5 and the diffuse interface heat conduction method developed in chap. 6, we finally applied the coupled diffuse interface method on a *real world engineering application* of the fluid flow and heat transfer in an open cell metal foam. This is an indicative example for the group of cellular solids and complex microstructures – one main area of application for phase field methods.

Simulations were performed for artificial counterparts of the foam structures used in the experiments, cf. chap. 7. The results in terms of pressure loss and heat transfer data are in good agreement with the measurements. Furthermore, the results provide vast information, which permits a detailed view on pore scale features of flow and temperature. Concluding we can determine that the derived methods are able to recover the integral performance values of fluid flow and heat transfer in complex foam-structures, with a satisfying and reasonable degree of accuracy.

---

[1]interface width per radius of curvature

Appendix A

Foam characterization

A.1 Pore diameter

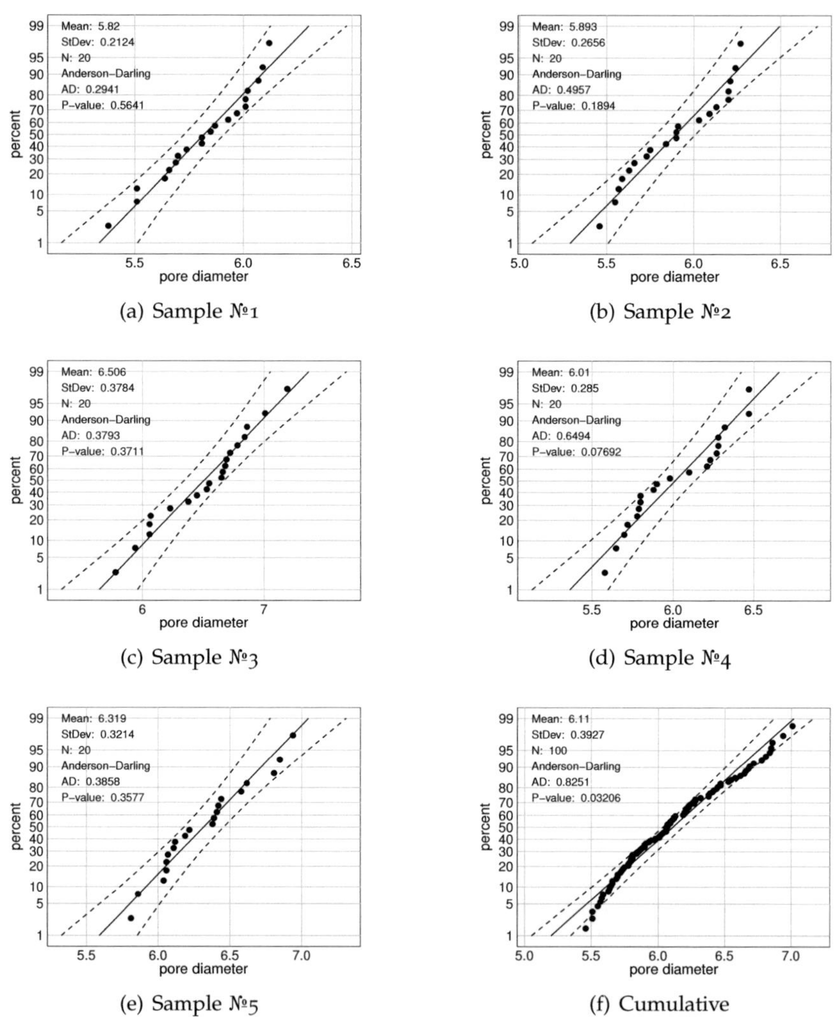

Figure A.1: Normal probability plots of pore diameter, (a)-(e) of the individual 10 ppi aluminium samples 1-5 and (f) cumultative normal probability plot.

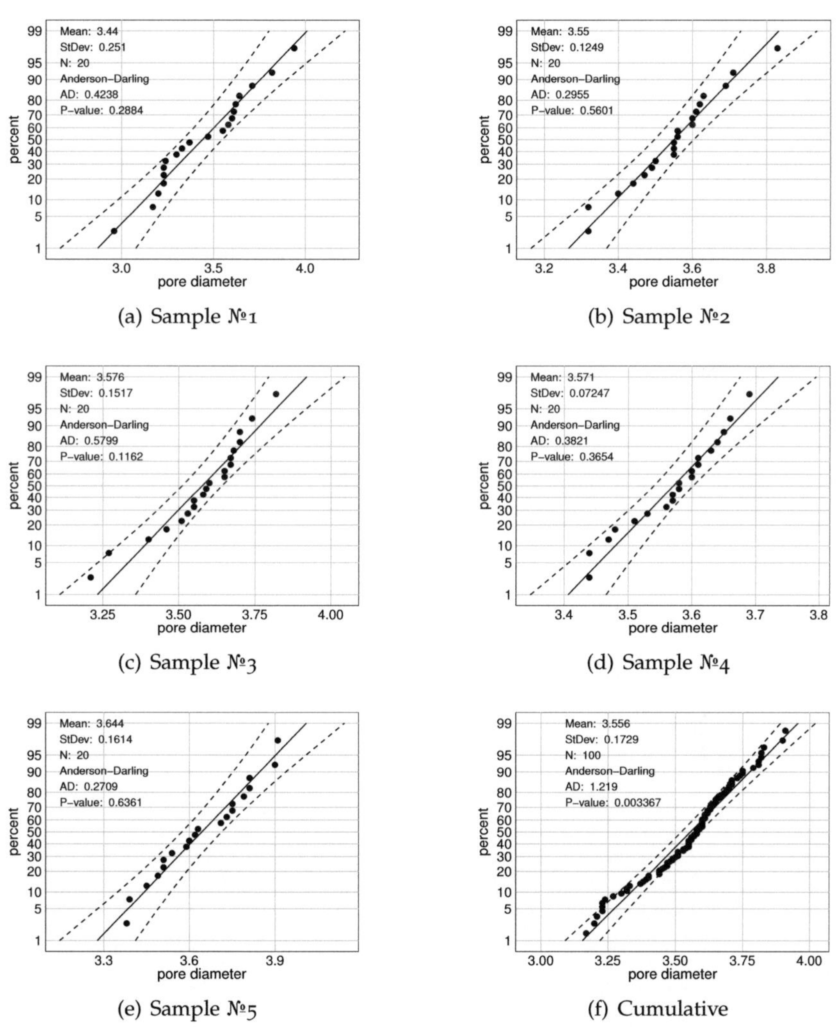

Figure A.2: Normal probability plots of pore diameter, (a)-(e) of the individual 20 ppi aluminium samples 1-5 and (f) cumultative normal probability plot.

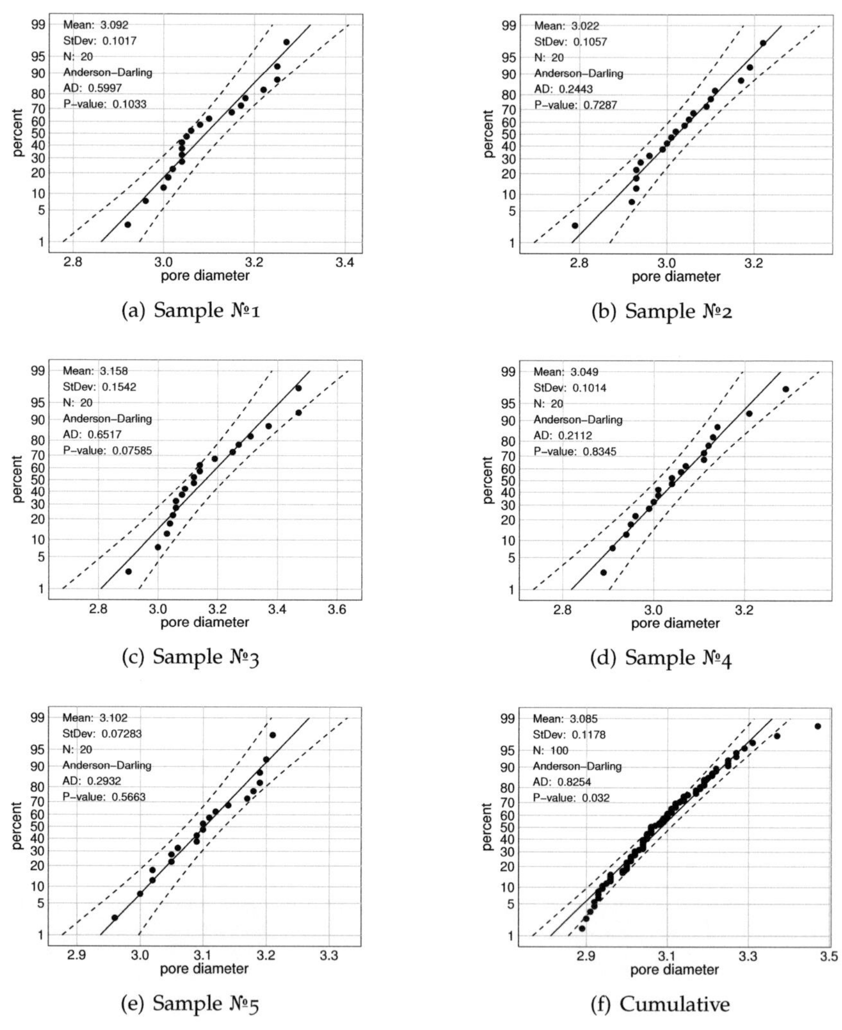

Figure A.3: Normal probability plots of pore diameter, (a)-(e) of the individual 30 ppi aluminium samples 1-5 and (f) cumultative normal probability plot.

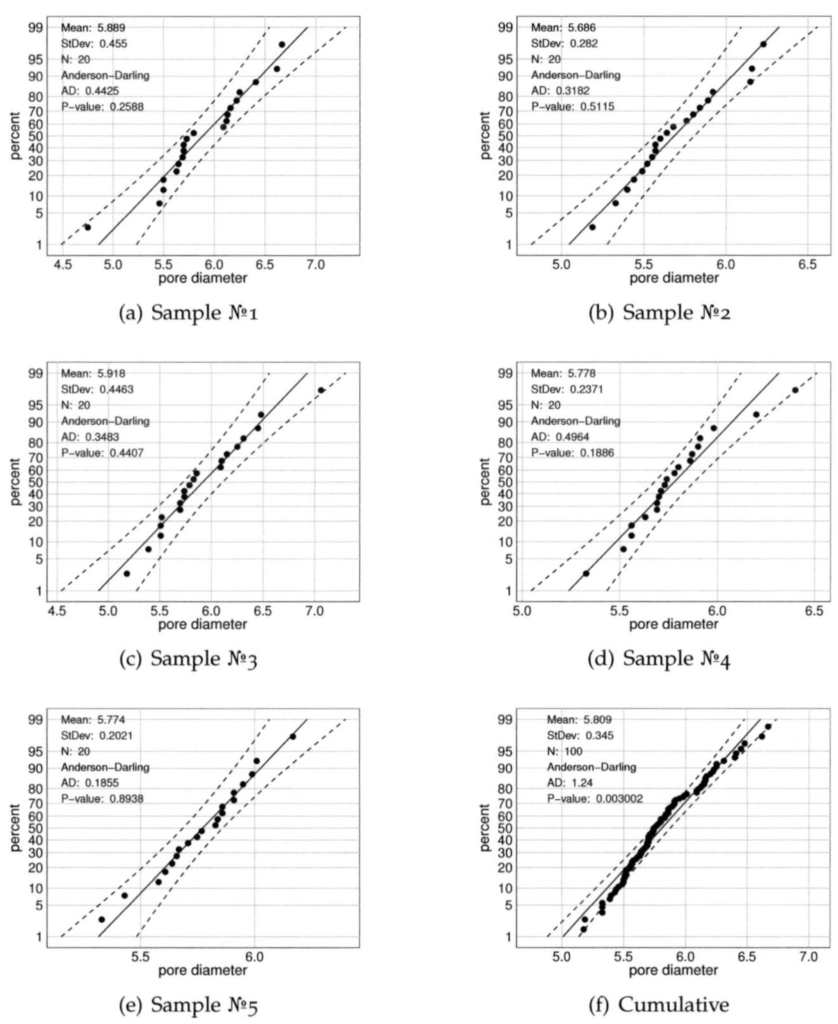

Figure A.4: Normal probability plots of pore diameter, (a)-(e) of the individual 10 ppi copper samples 1-5 and (f) cumultative normal probability plot.

A.2 Face diameter

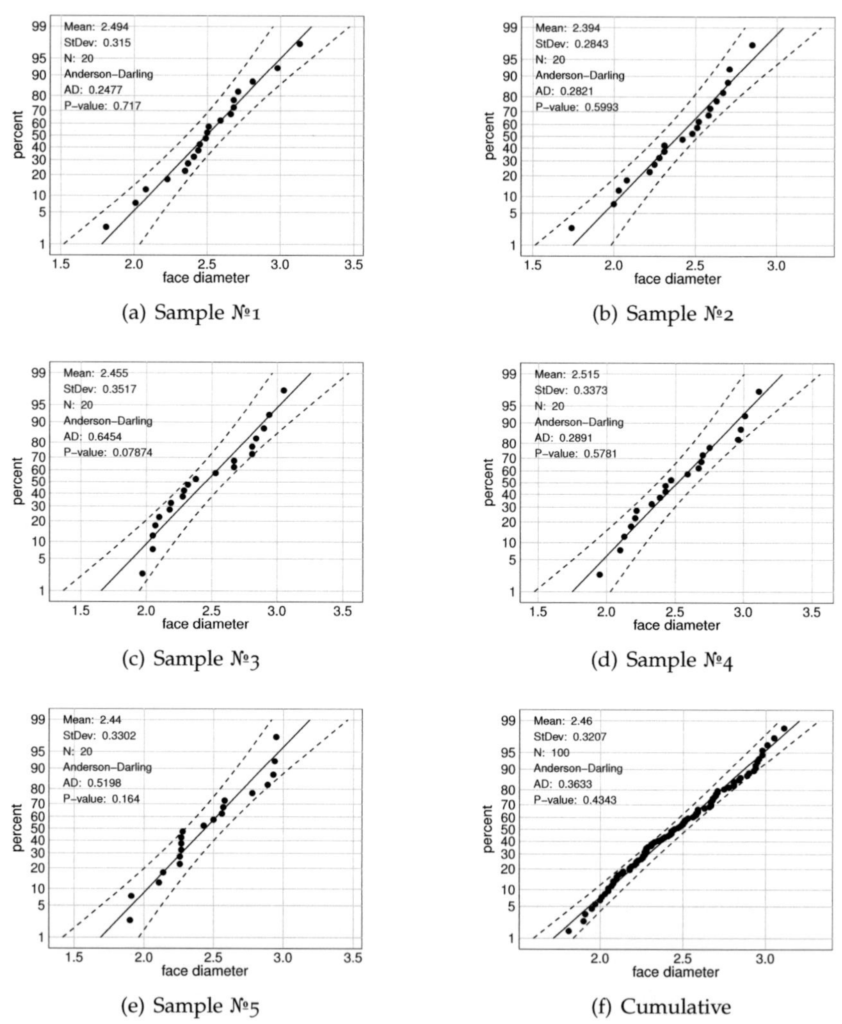

(a) Sample №1

(b) Sample №2

(c) Sample №3

(d) Sample №4

(e) Sample №5

(f) Cumulative

Figure A.5: Normal probability plots of face diameter, (a)-(e) of the individual 10 ppi aluminium samples 1-5 and (f) cumultative normal probability plot.

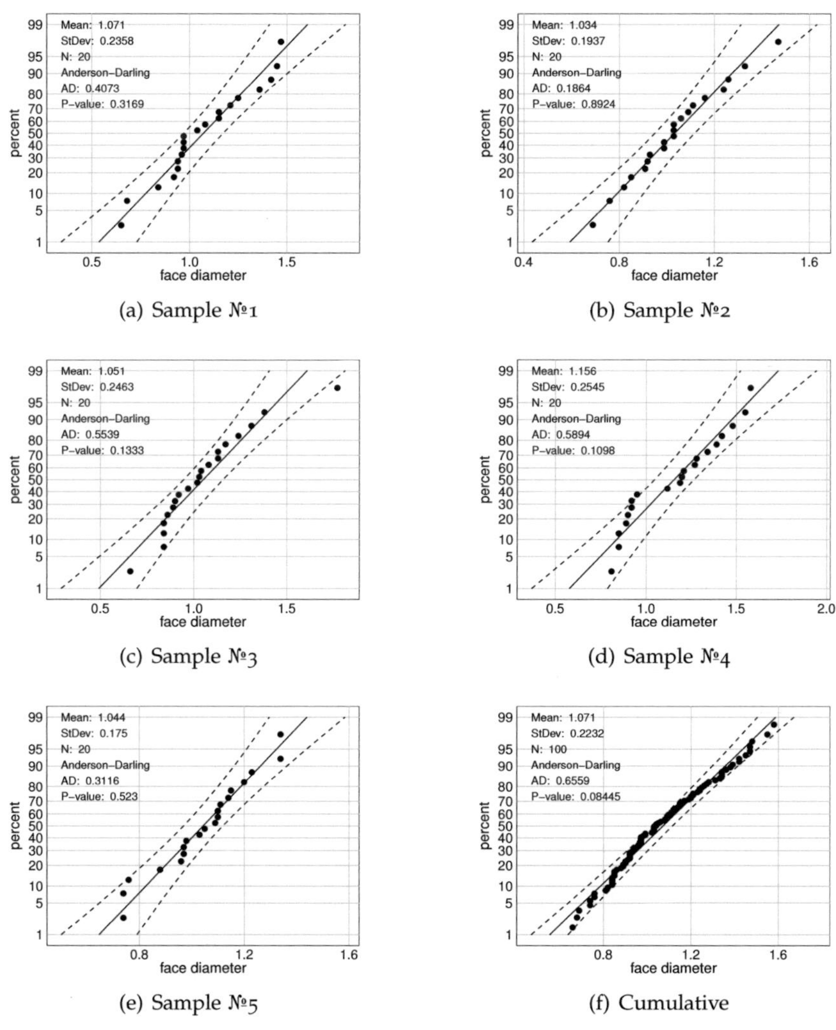

(a) Sample №1

(b) Sample №2

(c) Sample №3

(d) Sample №4

(e) Sample №5

(f) Cumulative

Figure A.6: Normal probability plots of face diameter, (a)-(e) of the individual 20 ppi aluminium samples 1-5 and (f) cumultative normal probability plot.

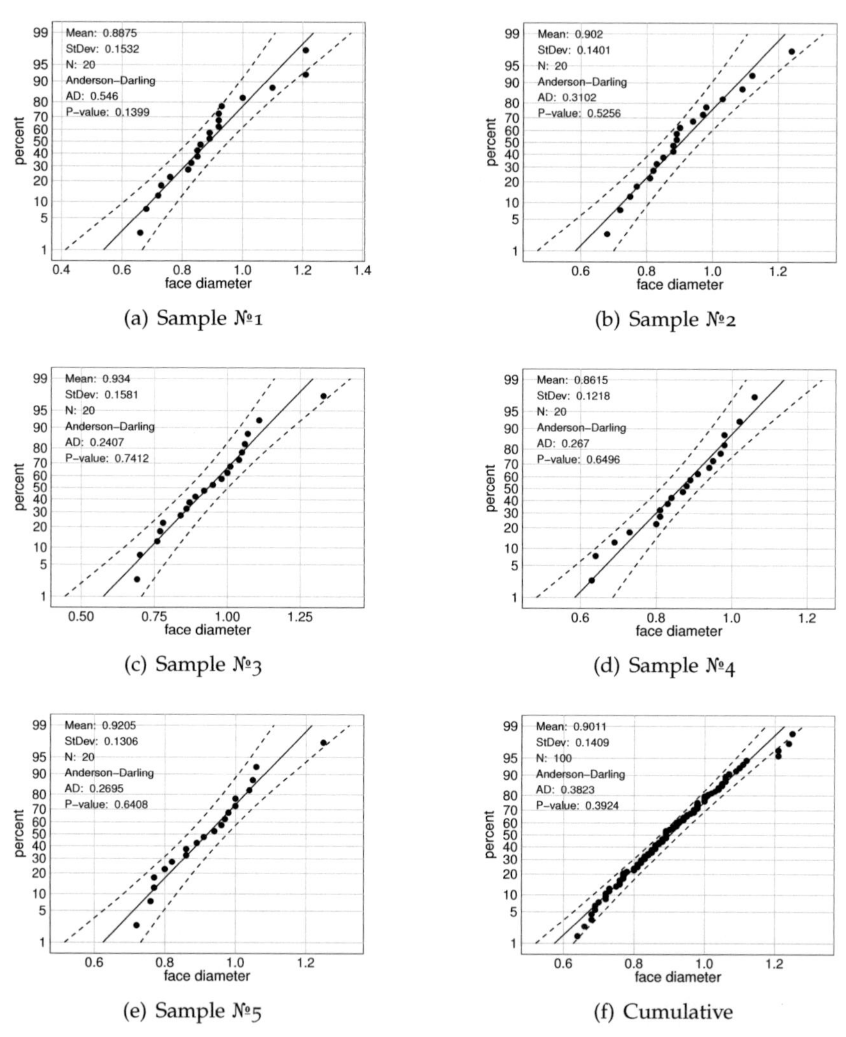

Figure A.7: Normal probability plots of face diameter, (a)-(e) of the individual 30 ppi aluminium samples 1-5 and (f) cumultative normal probability plot.

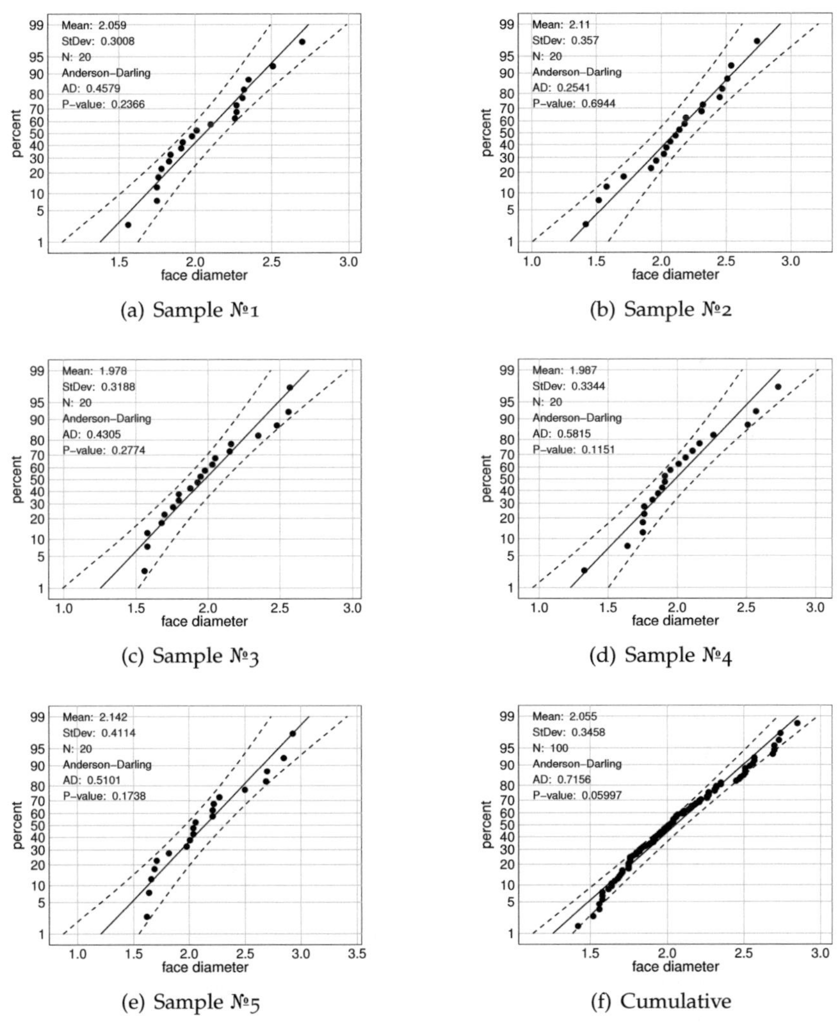

(a) Sample №1

(b) Sample №2

(c) Sample №3

(d) Sample №4

(e) Sample №5

(f) Cumulative

Figure A.8: Normal probability plots of face diameter, (a)-(e) of the individual 10 ppi copper samples 1-5 and (f) cumultative normal probability plot.

A.3 Edge thickness

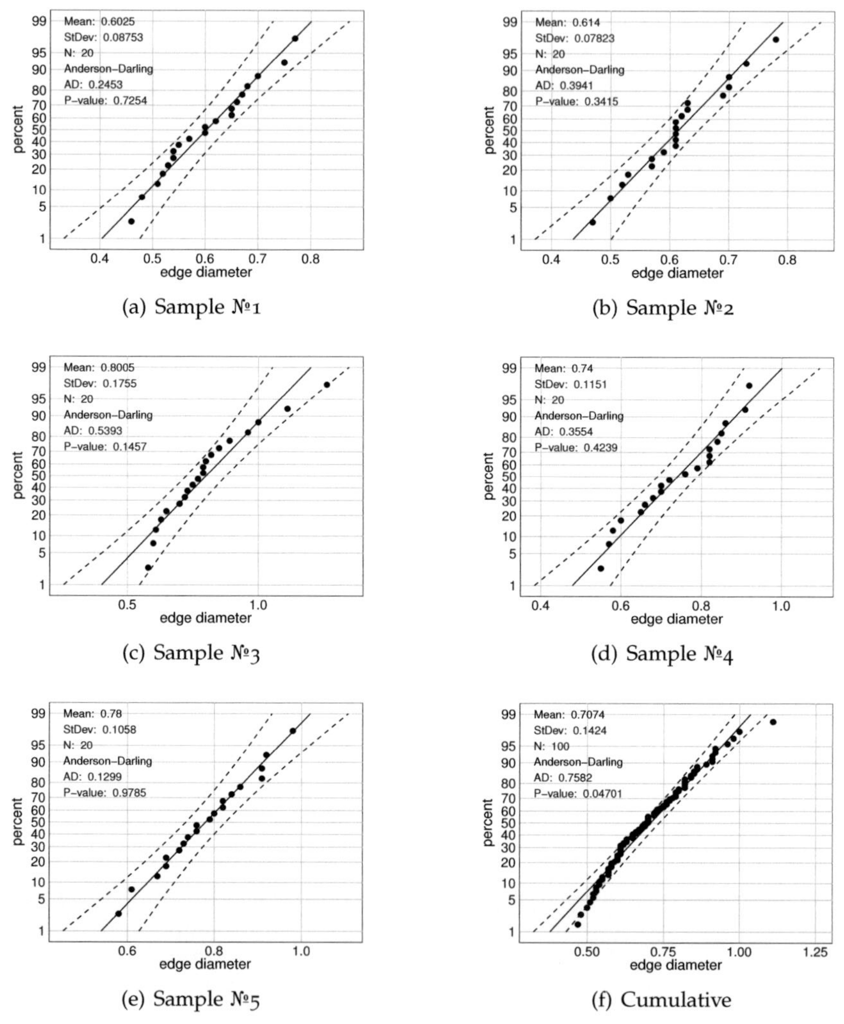

Figure A.9: Normal probability plots of edge thickness, (a)-(e) of the individual 10 ppi aluminium samples 1-5 and (f) cumultative normal probability plot.

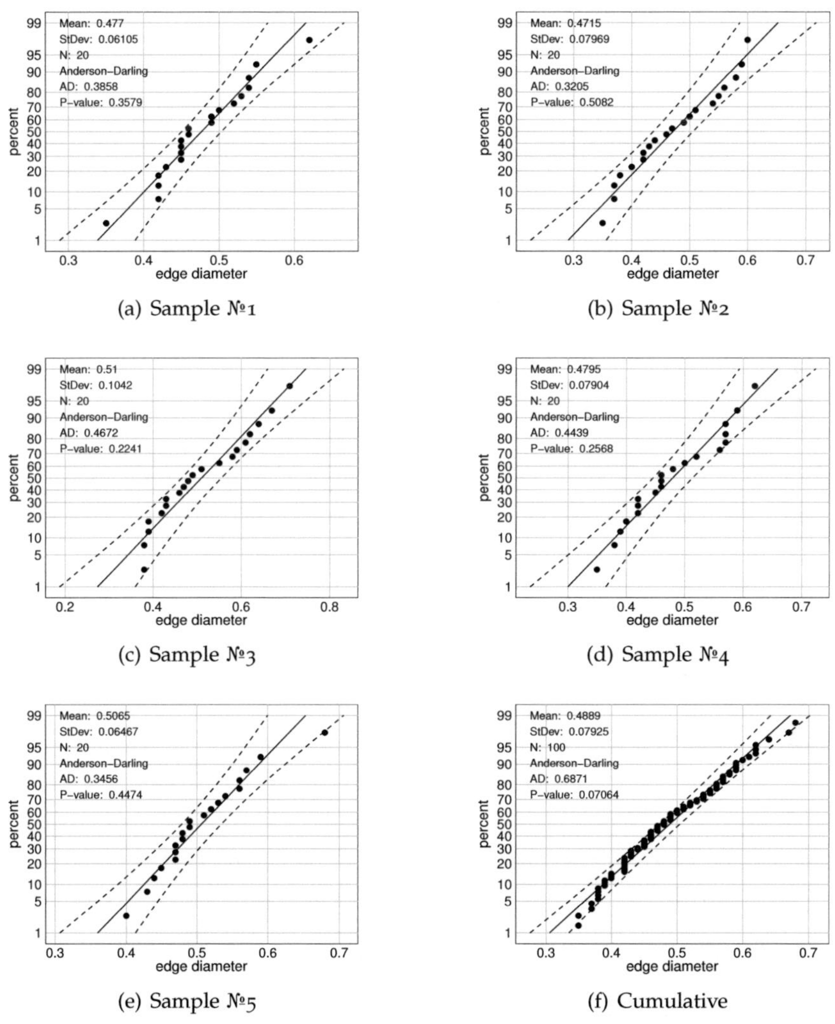

(a) Sample №1

(b) Sample №2

(c) Sample №3

(d) Sample №4

(e) Sample №5

(f) Cumulative

Figure A.10: Normal probability plots of edge thickness, (a)-(e) of the individual 20 ppi aluminium samples 1-5 and (f) cumultative normal probability plot.

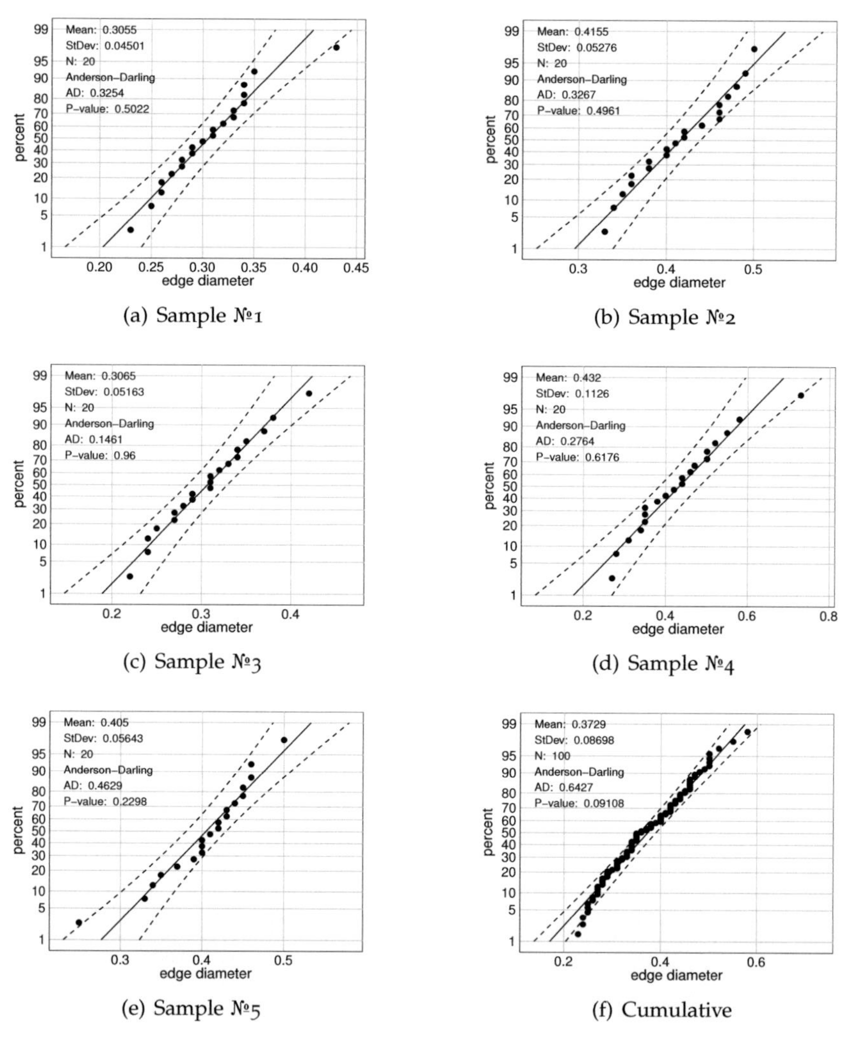

Figure A.11: Normal probability plots of edge thickness, (a)-(e) of the individual 30 ppi aluminium samples 1-5 and (f) cumultative normal probability plot.

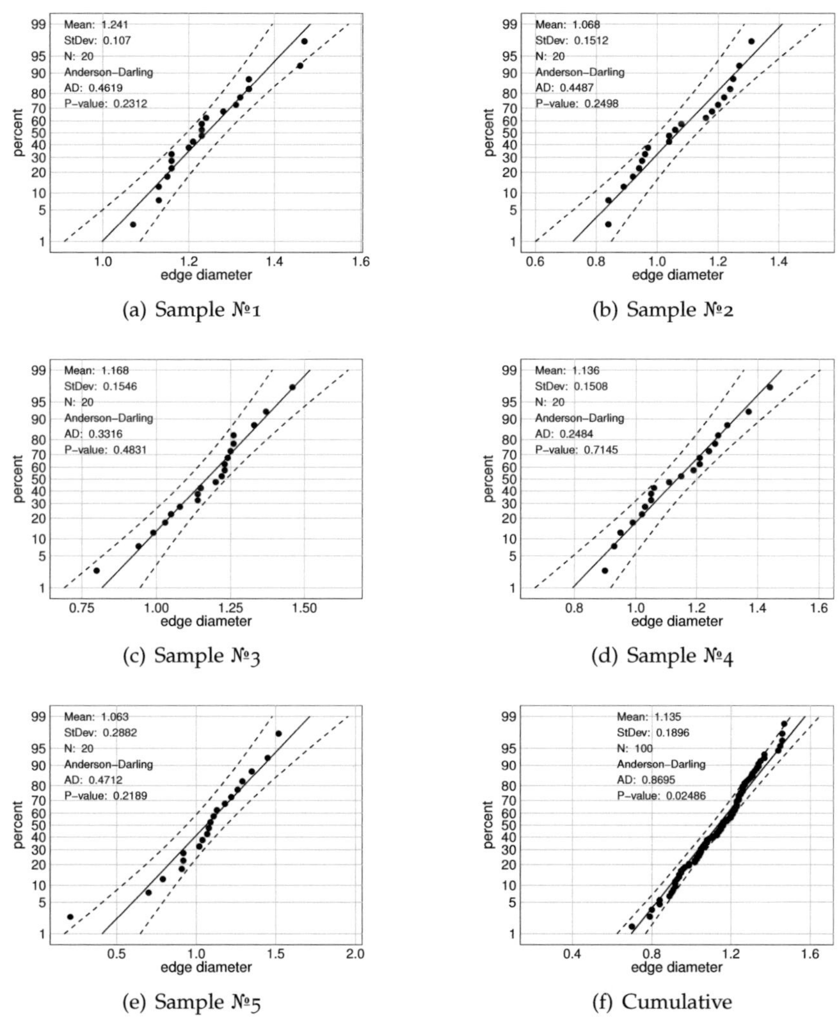

Figure A.12: Normal probability plots of edge thickness, (a)-(e) of the individual 10 ppi copper samples 1-5 and (f) cumultative normal probability plot.

Appendix B

Relative velocity profiles

B.1 10 ppi aluminium foam

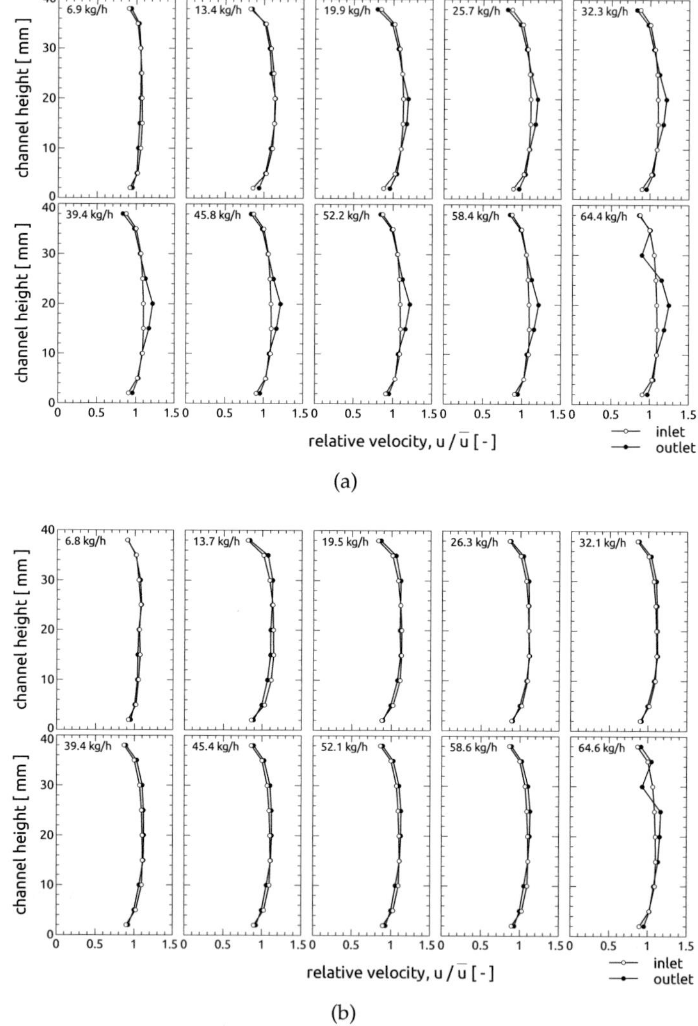

Figure B.1: Relative velocity $u/\bar{u}$ profiles for 10 ppi aluminium foam of (a) 20 mm and (b) 100 mm length, for different mass flows respectively.

B.2 20 ppi aluminium foam

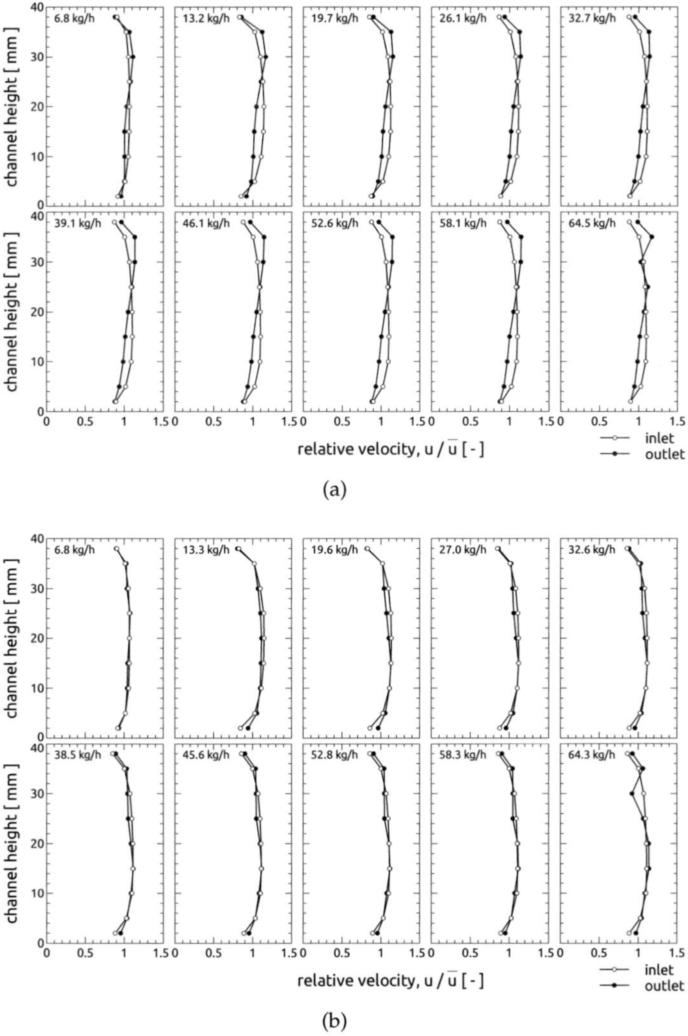

Figure B.2: Relative velocity $u/\bar{u}$ profiles for 20 ppi aluminium foam of (a) 20 mm and (b) 100 mm length, for different mass flows respectively.

B.3 30 ppi aluminium foam

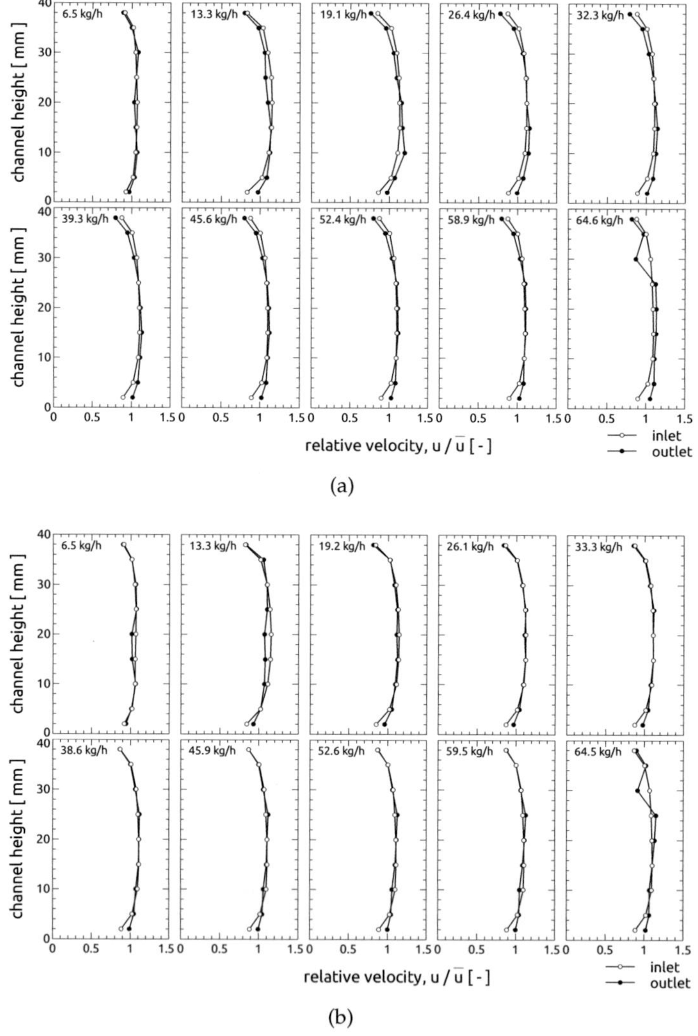

Figure B.3: Relative velocity $u/\bar{u}$ profiles for 30 ppi aluminium foam of (a) 20 mm and (b) 100 mm length, for different mass flows respectively.

B.4 10 ppi copper foam

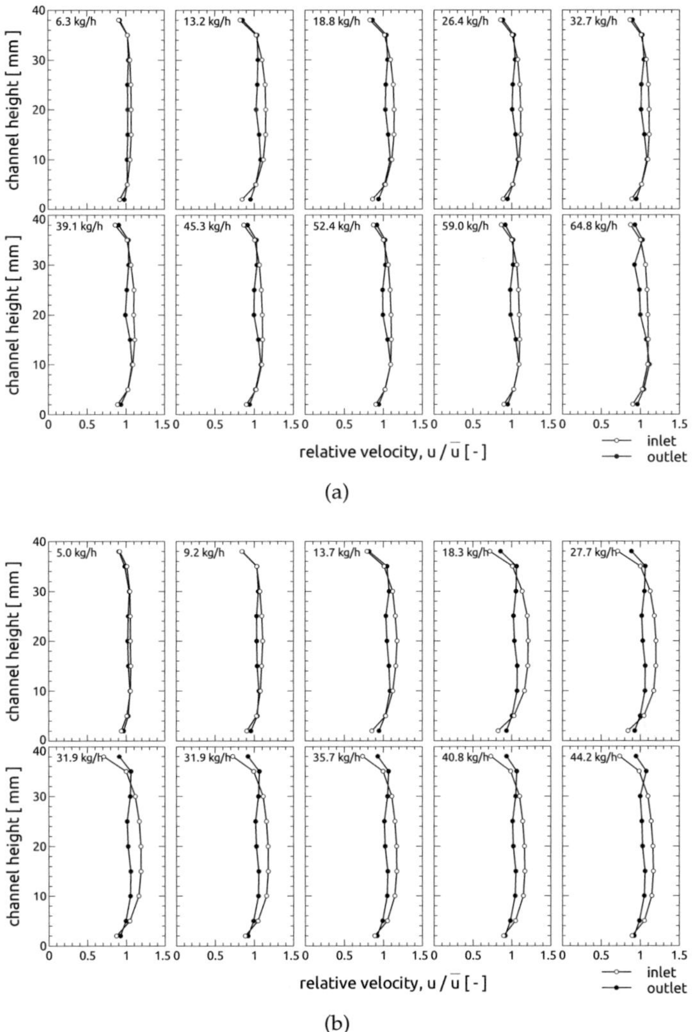

Figure B.4: Relative velocity $u/\bar{u}$ profiles for 10 ppi copper foam of (a) 20 mm and (b) 100 mm length, for different mass flows respectively.

Appendix C

Massflow distributions

C.1 10ppi aluminium foam

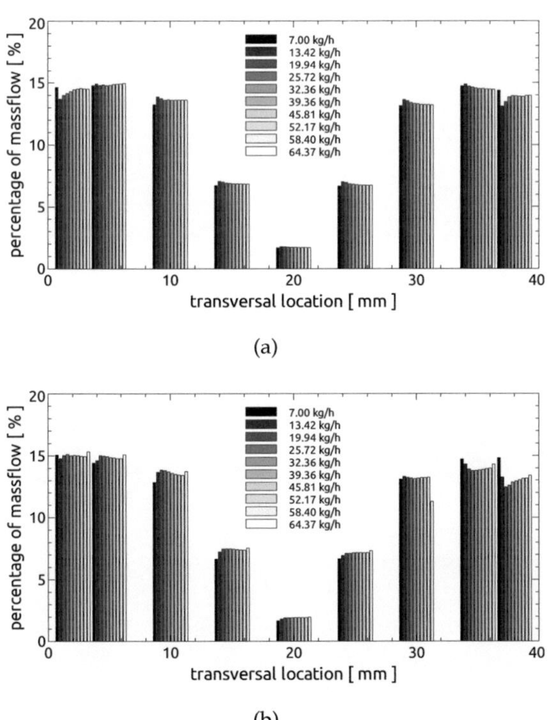

(a)

(b)

Figure C.1: Massflow distribution along transversal direction for 10 ppi aluminium foam of 20 mm length at (a) inlet and (b) outlet.

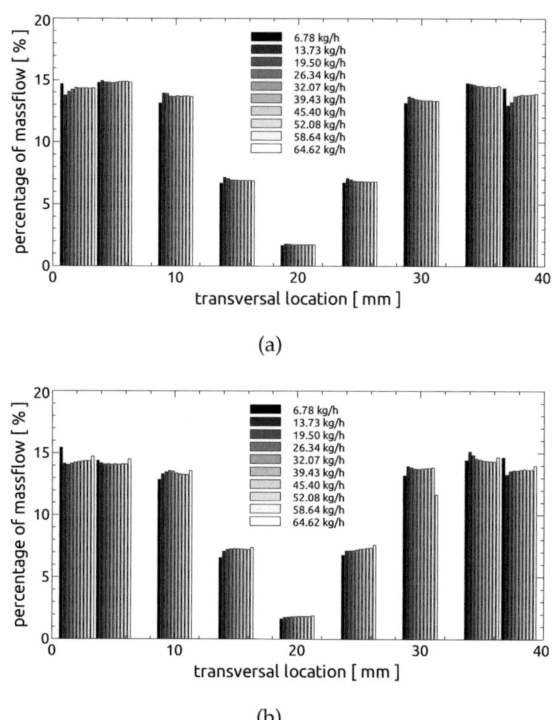

(a)

(b)

Figure C.2: Massflow distribution along transversal direction for 10 ppi aluminium foam of 100 mm length at (a) inlet and (b) outlet.

C.2 20ppi aluminium foam

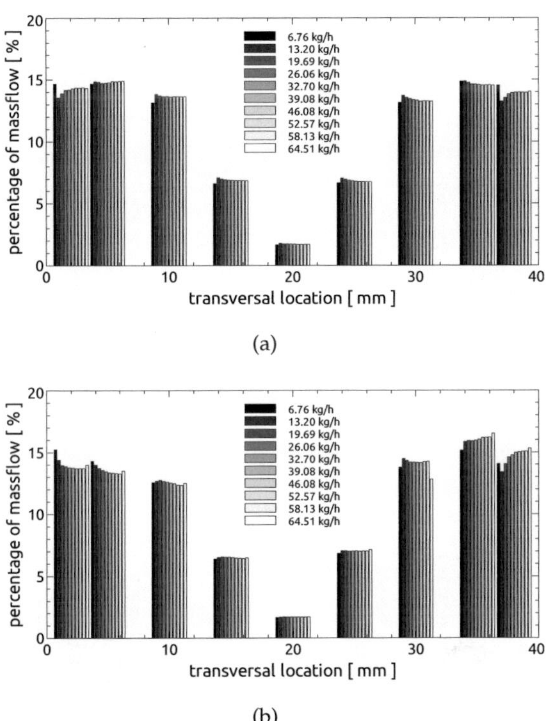

(a)

(b)

Figure C.3: Massflow distribution along transversal direction for 20 ppi aluminium foam of 20 mm length at (a) inlet and (b) outlet.

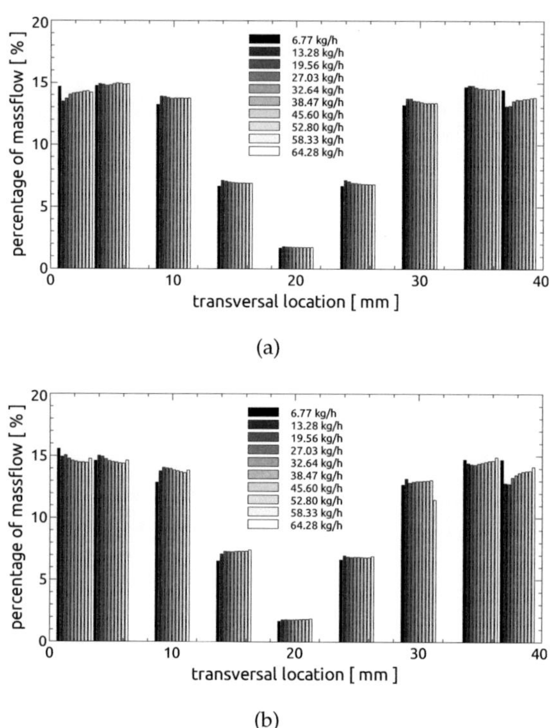

(a)

(b)

Figure C.4: Massflow distribution along transversal direction for 20 ppi aluminium foam of 100 mm length at (a) inlet and (b) outlet.

C.3 30ppi aluminium foam

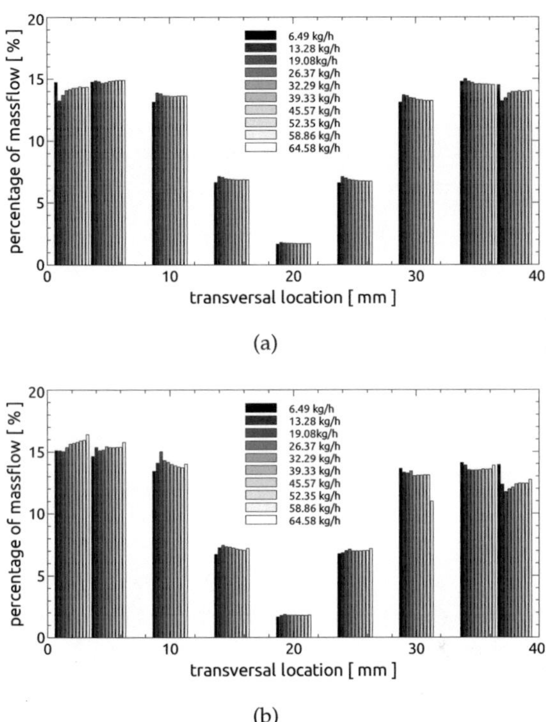

(a)

(b)

Figure C.5: Massflow distribution along transversal direction for 30 ppi aluminium foam of 20 mm length at (a) inlet and (b) outlet.

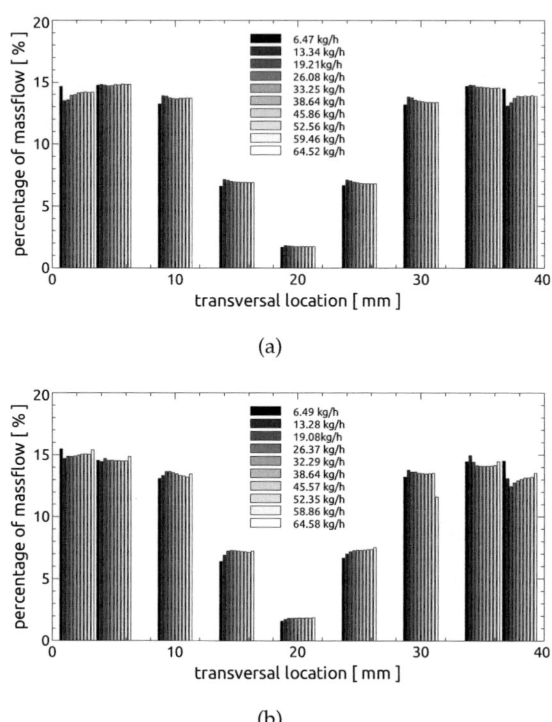

(a)

(b)

Figure C.6: Massflow distribution along transversal direction for 30 ppi aluminium foam of 100 mm length at (a) inlet and (b) outlet.

C.4 10ppi copper foam

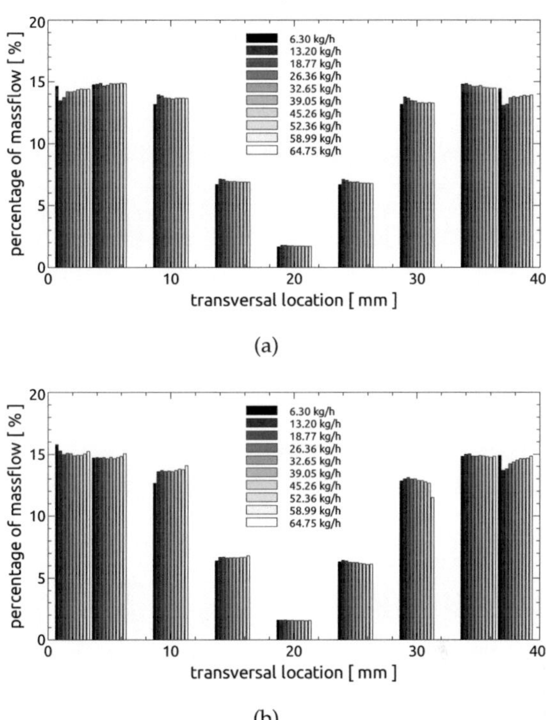

(a)

(b)

Figure C.7: Massflow distribution along transversal direction for 30 ppi aluminium foam of 20 mm length at (a) inlet and (b) outlet.

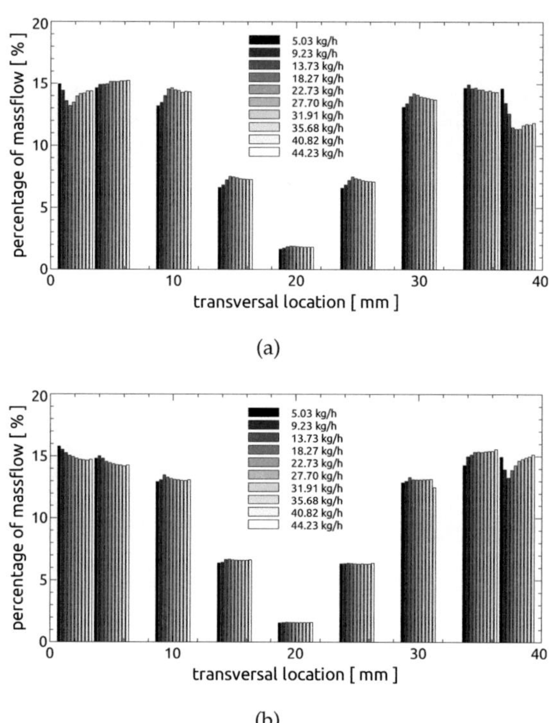

(a)

(b)

Figure C.8: Massflow distribution along transversal direction for 30 ppi aluminium foam of 100 mm length at (a) inlet and (b) outlet.

Appendix D

Parameters of the MRT lattice Boltzmann method

The full transformation matrices, inverse transformation matrices and relaxation parameters for the D2Q9 and D3Q19 MRT lattice Boltzmann model used in PACE3D are listed below.

D.1 D2Q9 multiple relaxation time modell

D2Q9 transformation matrix

$$
\mathcal{M} =
\begin{pmatrix}
1 & 1 & 1 & 1 & 1 & 1 & 1 & 1 & 1 \\
-4 & -1 & -1 & -1 & -1 & 2 & 2 & 2 & 2 \\
4 & -2 & -2 & -2 & -2 & 1 & 1 & 1 & 1 \\
0 & 0 & 0 & -1 & 1 & -1 & 1 & 1 & -1 \\
0 & 0 & 0 & 2 & -2 & -1 & 1 & 1 & -1 \\
0 & -1 & 1 & 0 & 0 & -1 & 1 & -1 & 1 \\
0 & 2 & -2 & 0 & 0 & -1 & 1 & -1 & 1 \\
0 & -1 & -1 & 1 & 1 & 0 & 0 & 0 & 0 \\
0 & 0 & 0 & 0 & 0 & 1 & 1 & -1 & -1
\end{pmatrix}
\tag{D.1}
$$

D2Q9 inverse transformation matrix

$$
\mathcal{M}^{-1} =
\begin{pmatrix}
\frac{1}{9} & -\frac{1}{9} & \frac{1}{9} & 0 & 0 & 0 & 0 & 0 & 0 \\
\frac{1}{9} & -\frac{1}{36} & -\frac{1}{18} & 0 & 0 & -\frac{1}{6} & \frac{1}{6} & -\frac{1}{4} & 0 \\
\frac{1}{9} & -\frac{1}{36} & -\frac{1}{18} & 0 & 0 & \frac{1}{6} & -\frac{1}{6} & -\frac{1}{4} & 0 \\
\frac{1}{9} & -\frac{1}{36} & \frac{1}{18} & -\frac{1}{6} & \frac{1}{6} & 0 & 0 & \frac{1}{4} & 0 \\
\frac{1}{9} & -\frac{1}{36} & -\frac{1}{18} & \frac{1}{6} & -\frac{1}{6} & 0 & 0 & \frac{1}{4} & 0 \\
\frac{1}{9} & \frac{1}{18} & \frac{1}{36} & -\frac{1}{6} & -\frac{1}{12} & -\frac{1}{6} & -\frac{1}{12} & 0 & \frac{1}{4} \\
\frac{1}{9} & \frac{1}{18} & \frac{1}{36} & \frac{1}{6} & \frac{1}{12} & \frac{1}{6} & \frac{1}{12} & 0 & \frac{1}{4} \\
\frac{1}{9} & \frac{1}{18} & \frac{1}{36} & \frac{1}{6} & \frac{1}{12} & -\frac{1}{6} & -\frac{1}{12} & 0 & \frac{1}{4} \\
\frac{1}{9} & \frac{1}{18} & \frac{1}{36} & -\frac{1}{6} & -\frac{1}{12} & \frac{1}{6} & \frac{1}{12} & 0 & -\frac{1}{4}
\end{pmatrix}
\tag{D.2}
$$

D2Q9 relaxation parameters

$$
s = \begin{pmatrix} 0.0 & 1.63 & 1.14 & 0.0 & 1.92 & 0.0 & 1.92 & 1.99 & 1.99 \end{pmatrix}
\tag{D.3}
$$

D.2 D3Q19 multiple relaxation time model

D3Q19 transformation matrix

$$\mathcal{M} = \begin{pmatrix}
1 & 1 & 1 & 1 & 1 & 1 & 1 & 1 & 1 & 1 & 1 & 1 & 1 & 1 & 1 & 1 & 1 & 1 & 1 \\
-30 & -11 & -11 & -11 & -11 & -11 & -11 & 8 & 8 & 8 & 8 & 8 & 8 & 8 & 8 & 8 & 8 & 8 & 8 \\
12 & -4 & -4 & -4 & -4 & -4 & -4 & 1 & 1 & 1 & 1 & 1 & 1 & 1 & 1 & 1 & 1 & 1 & 1 \\
0 & -1 & 1 & 0 & 0 & 0 & 0 & -1 & 1 & -1 & 1 & -1 & 1 & -1 & 1 & 0 & 0 & 0 & 0 \\
0 & 4 & -4 & 0 & 0 & 0 & 0 & -1 & 1 & -1 & 1 & -1 & 1 & -1 & 1 & 0 & 0 & 0 & 0 \\
0 & 0 & 0 & -1 & 1 & 0 & 0 & 0 & 0 & 0 & 0 & -1 & 1 & 1 & -1 & -1 & 1 & -1 & 1 \\
0 & 0 & 0 & 4 & -4 & 0 & 0 & 0 & 0 & 0 & 0 & -1 & 1 & 1 & -1 & -1 & 1 & -1 & 1 \\
0 & 0 & 0 & 0 & 0 & -1 & 1 & -1 & 1 & 1 & -1 & 0 & 0 & 0 & 0 & -1 & 1 & 1 & -1 \\
0 & 0 & 0 & 0 & 0 & 4 & -4 & -1 & 1 & 1 & -1 & 0 & 0 & 0 & 0 & -1 & 1 & 1 & -1 \\
0 & 2 & 2 & -1 & -1 & -1 & -1 & 1 & 1 & 1 & 1 & 1 & 1 & 1 & 1 & -2 & -2 & -2 & -2 \\
0 & -4 & -4 & 2 & 2 & 2 & 2 & 1 & 1 & 1 & 1 & 1 & 1 & 1 & 1 & -2 & -2 & -2 & -2 \\
0 & 0 & 0 & 1 & 1 & -1 & -1 & -1 & -1 & -1 & -1 & 1 & 1 & 1 & 1 & 0 & 0 & 0 & 0 \\
0 & 0 & 0 & -2 & -2 & 2 & 2 & -1 & -1 & -1 & -1 & 1 & 1 & 1 & 1 & 0 & 0 & 0 & 0 \\
0 & 0 & 0 & 0 & 0 & 0 & 0 & 0 & 0 & 0 & 0 & 1 & 1 & -1 & -1 & 0 & 0 & 0 & 0 \\
0 & 0 & 0 & 0 & 0 & 0 & 0 & 0 & 0 & 0 & 0 & 0 & 0 & 0 & 0 & 1 & 1 & -1 & -1 \\
0 & 0 & 0 & 0 & 0 & 0 & 0 & 1 & 1 & -1 & -1 & 0 & 0 & 0 & 0 & 0 & 0 & 0 & 0 \\
0 & 0 & 0 & 0 & 0 & 0 & 0 & 1 & -1 & 1 & -1 & -1 & 1 & -1 & 1 & 0 & 0 & 0 & 0 \\
0 & 0 & 0 & 0 & 0 & 0 & 0 & 0 & 0 & 0 & 0 & 1 & -1 & -1 & 1 & -1 & 1 & -1 & 1 \\
0 & 0 & 0 & 0 & 0 & 0 & 0 & -1 & 1 & 1 & -1 & 0 & 0 & 0 & 0 & 1 & -1 & -1 & 1
\end{pmatrix} \tag{D.4}$$

D3Q19 inverse transformation matrix

$$\mathcal{M}^{-1} = \begin{pmatrix}
\frac{1}{19} & -\frac{5}{399} & \frac{1}{21} & 0 & 0 & 0 & 0 & 0 & 0 & 0 & 0 & 0 & 0 & 0 & 0 & 0 & 0 & 0 & 0 \\
\frac{1}{19} & -\frac{11}{2394} & -\frac{1}{63} & -\frac{1}{10} & \frac{1}{10} & 0 & 0 & 0 & 0 & \frac{1}{18} & -\frac{1}{18} & 0 & 0 & 0 & 0 & 0 & 0 & 0 & 0 \\
\frac{1}{19} & -\frac{11}{2394} & -\frac{1}{63} & \frac{1}{10} & -\frac{1}{10} & 0 & 0 & 0 & 0 & \frac{1}{18} & -\frac{1}{18} & 0 & 0 & 0 & 0 & 0 & 0 & 0 & 0 \\
\frac{1}{19} & -\frac{11}{2394} & -\frac{1}{63} & 0 & 0 & -\frac{1}{10} & \frac{1}{10} & 0 & 0 & -\frac{1}{36} & \frac{1}{36} & \frac{1}{12} & -\frac{1}{12} & 0 & 0 & 0 & 0 & 0 & 0 \\
\frac{1}{19} & -\frac{11}{2394} & -\frac{1}{63} & 0 & 0 & \frac{1}{10} & -\frac{1}{10} & 0 & 0 & -\frac{1}{36} & \frac{1}{36} & \frac{1}{12} & -\frac{1}{12} & 0 & 0 & 0 & 0 & 0 & 0 \\
\frac{1}{19} & -\frac{11}{2394} & -\frac{1}{63} & 0 & 0 & 0 & 0 & -\frac{1}{10} & \frac{1}{10} & -\frac{1}{36} & \frac{1}{36} & -\frac{1}{12} & \frac{1}{12} & 0 & 0 & 0 & 0 & 0 & 0 \\
\frac{1}{19} & -\frac{11}{2394} & -\frac{1}{63} & 0 & 0 & 0 & 0 & \frac{1}{10} & -\frac{1}{10} & -\frac{1}{36} & \frac{1}{36} & -\frac{1}{12} & \frac{1}{12} & 0 & 0 & 0 & 0 & 0 & 0 \\
\frac{1}{19} & \frac{4}{1197} & \frac{1}{252} & -\frac{1}{10} & -\frac{1}{40} & 0 & 0 & -\frac{1}{10} & -\frac{1}{40} & \frac{1}{36} & \frac{1}{72} & -\frac{1}{12} & -\frac{1}{24} & 0 & 0 & \frac{1}{4} & \frac{1}{8} & 0 & -\frac{1}{8} \\
\frac{1}{19} & \frac{4}{1197} & \frac{1}{252} & \frac{1}{10} & \frac{1}{40} & 0 & 0 & \frac{1}{10} & \frac{1}{40} & \frac{1}{36} & \frac{1}{72} & -\frac{1}{12} & -\frac{1}{24} & 0 & 0 & \frac{1}{4} & -\frac{1}{8} & 0 & \frac{1}{8} \\
\frac{1}{19} & \frac{4}{1197} & \frac{1}{252} & -\frac{1}{10} & -\frac{1}{40} & 0 & 0 & \frac{1}{10} & \frac{1}{40} & \frac{1}{36} & \frac{1}{72} & -\frac{1}{12} & -\frac{1}{24} & 0 & 0 & -\frac{1}{4} & \frac{1}{8} & 0 & \frac{1}{8} \\
\frac{1}{19} & \frac{4}{1197} & \frac{1}{252} & \frac{1}{10} & \frac{1}{40} & 0 & 0 & -\frac{1}{10} & -\frac{1}{40} & \frac{1}{36} & \frac{1}{72} & -\frac{1}{12} & -\frac{1}{24} & 0 & 0 & -\frac{1}{4} & -\frac{1}{8} & 0 & -\frac{1}{8} \\
\frac{1}{19} & \frac{4}{1197} & \frac{1}{252} & -\frac{1}{10} & -\frac{1}{40} & -\frac{1}{10} & -\frac{1}{40} & 0 & 0 & \frac{1}{36} & \frac{1}{72} & \frac{1}{12} & \frac{1}{24} & \frac{1}{4} & 0 & 0 & -\frac{1}{8} & \frac{1}{8} & 0 \\
\frac{1}{19} & \frac{4}{1197} & \frac{1}{252} & \frac{1}{10} & \frac{1}{40} & \frac{1}{10} & \frac{1}{40} & 0 & 0 & \frac{1}{36} & \frac{1}{72} & \frac{1}{12} & \frac{1}{24} & \frac{1}{4} & 0 & 0 & \frac{1}{8} & -\frac{1}{8} & 0 \\
\frac{1}{19} & \frac{4}{1197} & \frac{1}{252} & -\frac{1}{10} & -\frac{1}{40} & \frac{1}{10} & \frac{1}{40} & 0 & 0 & \frac{1}{36} & \frac{1}{72} & \frac{1}{12} & \frac{1}{24} & -\frac{1}{4} & 0 & 0 & -\frac{1}{8} & -\frac{1}{8} & 0 \\
\frac{1}{19} & \frac{4}{1197} & \frac{1}{252} & \frac{1}{10} & \frac{1}{40} & -\frac{1}{10} & -\frac{1}{40} & 0 & 0 & \frac{1}{36} & \frac{1}{72} & \frac{1}{12} & \frac{1}{24} & -\frac{1}{4} & 0 & 0 & \frac{1}{8} & \frac{1}{8} & 0 \\
\frac{1}{19} & \frac{4}{1197} & \frac{1}{252} & 0 & 0 & -\frac{1}{10} & -\frac{1}{40} & -\frac{1}{10} & -\frac{1}{40} & -\frac{1}{18} & -\frac{1}{36} & 0 & 0 & 0 & \frac{1}{4} & 0 & 0 & -\frac{1}{8} & \frac{1}{8} \\
\frac{1}{19} & \frac{4}{1197} & \frac{1}{252} & 0 & 0 & \frac{1}{10} & \frac{1}{40} & \frac{1}{10} & \frac{1}{40} & -\frac{1}{18} & -\frac{1}{36} & 0 & 0 & 0 & \frac{1}{4} & 0 & 0 & \frac{1}{8} & -\frac{1}{8} \\
\frac{1}{19} & \frac{4}{1197} & \frac{1}{252} & 0 & 0 & -\frac{1}{10} & -\frac{1}{40} & \frac{1}{10} & \frac{1}{40} & -\frac{1}{18} & -\frac{1}{36} & 0 & 0 & 0 & -\frac{1}{4} & 0 & 0 & -\frac{1}{8} & -\frac{1}{8} \\
\frac{1}{19} & \frac{4}{1197} & \frac{1}{252} & 0 & 0 & \frac{1}{10} & \frac{1}{40} & -\frac{1}{10} & -\frac{1}{40} & -\frac{1}{18} & -\frac{1}{36} & 0 & 0 & 0 & -\frac{1}{4} & 0 & 0 & \frac{1}{8} & \frac{1}{8}
\end{pmatrix} \tag{D.5}$$

D3Q19 relaxation parameters

$$s = (\; 0.0 \quad 1.19 \quad 1.40 \quad 0.0 \quad 1.20 \quad 0.0 \quad 1.20 \quad 0.0 \quad 1.20 \quad 0.0 \quad 1.40 \quad 0.0 \quad 1.40 \quad 0.0 \quad 0.0 \quad 0.0 \quad 1.98 \quad 1.98 \quad 1.98 \;) \tag{D.6}$$

Appendix E

Parametrisation

E.1 Exemplary parametrisation of the lattice Boltzmann fluid flow solver

In this section the parametrisation of the lattice Boltzmann flow solver in PACE3D is shown according to sec. 5.9. For this, we consider the flow of air (kinematic viscosity $v_{(p)} = 1.8558 \times 10^{-5}\mathrm{m}^2/\mathrm{s}$) at superficial velocity of $u_{(p)} = 2\,\mathrm{m/s}$ in a cylindrical channel with a diameter of $l_{(p)} = 0.04\,\mathrm{m}$. Thus, the characteristic Reynolds number is given as

$$\mathrm{Re}_{(p)} = \frac{u_{(p)} \cdot l_{(p)}}{v_{(p)}} = \frac{2\mathrm{m/s} \cdot 0.04\mathrm{m}}{1.8558 \times 10^{-5}\mathrm{m}^2/\mathrm{s}} \approx 4312\,. \tag{E.1}$$

With a choice for the spatial resolution N_x and the magnitude of the velocity in lattice units $u_{(lb)}$ we can deduce the discrete spatial interval Δx and the discrete time interval Δt. For instance, using $N_x = 400$ and $u_{(lb)} = 0,05$ we get

$$\delta x = \frac{l_{(p)}}{N_x} = \frac{0.04\mathrm{m}}{400} = 0.0001\mathrm{m}\,, \tag{E.2}$$

and

$$\delta t = \frac{u_{(lb)}}{u_{(lb)}} \cdot \delta x = \frac{0.05}{2\mathrm{m/s}} \cdot 0.0001\mathrm{m} = 0.0000025\mathrm{s} \tag{E.3}$$

Using the discrete spatial and temporal interval δx and δt, we get the lattice viscosity $v_{(lb)}$ from

$$v_{(lb)} = v_{(p)}\frac{\delta t}{\delta x^2} = 1.8558 \times 10^{-5}\mathrm{m}^2/\mathrm{s} \cdot \frac{0.0000025\mathrm{s}}{(0.0001\mathrm{m})^2} = 0.0046377\,. \tag{E.4}$$

The relaxation parameter τ follows from eq. (5.21) as

$$\tau = 3 \cdot \nu_{(lb)} + 0.5 = 0.5139, \qquad\qquad (E.5)$$

whereas the required number of timesteps N_t to cover certain timespan Δt is

$$N_t = \frac{\Delta t}{\delta t}. \qquad\qquad (E.6)$$

```
#
# LBM SPECIFIC
#
FluidDynamics.LiquidPhases                      = (0.0,  1.0)
FluidDynamics.Gamma                             = 0.5
FluidDynamics.streakline                        = 0
FluidDynamics.traceline                         = 0
FluidDynamics.Gravity                           = (0.0,  0.0,  0.0)
FluidDynamics.Density                           = (1.0,  1.0)

FluidDynamics.LBM.KinematicViscosity            = (0.0,  0.0046377)
FluidDynamics.LBM.ReferenceDensity              =  1.0
FluidDynamics.LBM.K                             = (1.0,  1.0)

FluidDynamics.LBM.Store.Residuals               = 0
FluidDynamics.LBM.Store.VelocityMagnitude       = 1
FluidDynamics.LBM.Store.VelocityComponents      = 1

FluidDynamics.LBM.LatticeType                   = d3q19
FluidDynamics.LBM.Dynamics                      = 0
FluidDynamics.LBM.DeltaX                        = 0.000 1

FluidDynamics.LBM.Boundary.ClosedWall           = (0,0,0,0,0,0)
FluidDynamics.LBM.Boundary.Pressure             = (1,1,1,1,1,1)
FluidDynamics.LBM.InnerBoundary                 =  no_slip
FluidDynamics.LBM.RotationAxis                  = -1

FluidDynamics.LBM.Precalc                       = 0
FluidDynamics.LBM.Precalc.Steps                 = 0
FluidDynamics.LBM.Precalc.Threshold             = 0.000 01

FluidDynamics.LBM.Init                          = 1
FluidDynamics.LBM.Init.Velocity                 = ( 0.0, 0.0, 0.0 )

FluidDynamics.Boundary.FlowRate = [(0.0,  0.0,  0.0 ),  (0.0,  0.0,  0.0 ),\
                                   (0.0,  0.0,  0.0 ),  (0.0,  0.0,  0.0 ),\
                                   (0.0,  0.0,  0.05),  (0.0,  0.0,  0.05) ]

FluidDynamics.LBM.Boundary        = ( periodic,        periodic, \
                                      periodic,        periodic, \
                                      velocity_linear, pressure )
```

Listing E.1: Lattice Boltzmann specific control parameters

Finally, we can check the Reynolds number of the discrete system to be equal to the physical Reynolds number

$$\mathrm{Re}_{(lb)} = \frac{N_x \cdot u_{(lb)}}{\nu_{(lb)}} \qquad (E.7)$$

An exemplary listing of the concerning lattice Boltzmann specific control parameters is given in listing E.1, where the parameter

```
FluidDynamics.LBM.KinematicViscosity
```

represents the kinematic viscosities of all phases and velocities in lattice units at the boundaries are given using the parameter

```
FluidDynamics.Boundary.FlowRate.
```

E.2 Exemplary parametrisation of the finite difference heat transfer solver

In this section the parametrisation of the heat equation solver in PACE3D is shown according to sec. 6.10. We assume that the considered problem is fully discribed by a characteristic length $l_{\text{ref}} = 0.001\,\text{m}$, a characteristic density $\rho_{\text{ref}} = 1\,\text{kg/m}^3$, a characteristic temperature $T_{\text{ref}} = 300\,\text{K}$ and a characteristic time interval $t_{\text{ref}} = 0.0000025\,\text{s}$.

Starting from the physical values of e.g. aluminium for the specific heat capacity $c_{V,(p)} = 8.37 \times 10^2\,\text{J}/(\text{kg\,K})$, the volumetric heat capacity $C_{V,(p)} = 2.344 \times 10^6\,\text{J}/(\text{m}^3\,\text{K})$ and the thermal conductivity $k_{(p)} = 1.5 \times 10^2\,\text{W}/(\text{m\,K})$ we can derive the non-dimensional values according to eqns. 6.66 as

$$c_{V,(n)} = c_{V,(p)} \cdot \frac{t_{\text{ref}}^2 T_{\text{ref}}}{x_{\text{ref}}^2} = 8.37 \times 10^2\,\text{J}/(\text{kg\,K}) \cdot \frac{(0.0000025\,\text{s})^2 \cdot 300\,\text{K}}{(0.001\,\text{m})^2} = 1.569 \times 10^2$$

$$C_{V,(n)} = C_{V,(p)} \cdot \frac{t_{\text{ref}}^2 T_{\text{ref}}}{\rho_{\text{ref}} x_{\text{ref}}^2} = 2.344 \times 10^6\,\text{J}/(\text{m}^3\,\text{K}) \cdot \frac{(0.0000025\,\text{s})^2 \cdot 300\,\text{K}}{1\,\text{kg/m}^3 \cdot (0.001\,\text{m})^2} = 4.394 \times 10^5$$

$$k_{(n)} = k_{(p)} \cdot \frac{t_{\text{ref}}^3 T_{\text{ref}}}{\rho_{\text{ref}} x_{\text{ref}}^4} = 1.5 \times 10^2\,\text{W}/(\text{m\,K}) \cdot \frac{(0.0000025\,\text{s})^3 \cdot 300\,\text{K}}{1\,\text{kg/m}^3 \cdot (0.001\,\text{m})^4} = 7.031 \times 10^3 \,.$$

$$\text{(E.8)}$$

```
 1  #
 2  # THERMAL SPECIFICATION
 3  #
 4  Energy.Type          = 6
 5  Energy.Noise.Type    = 0
 6  Energy.CV.Type       = 0
 7  Energy.k.Type        = 0
 8  Energy.CV.CV         = 1.0
 9  Energy.CV.CVa        = ( 439425.        , 261.92 )
10  Energy.k.Ka          = (   7031.25      ,   1.172)
11  Energy.CV.fa         = (        5.05906,   1.   )
12
13  Energy.option        = diffusion+advection_upwind
14  Energy.scale         = 0
15  Energy.scale.vectorT = (1, 1)
16  Energy.Boundary      = ( isolate, isolate, \
17                           isolate, isolate, \
18                           isolate, isolate  )
19
20  Temperature          = 1.
```

Listing E.2: Heat equation specific control parameters

The relevant control parameters are given in listing E.2, where the parameter `Energy.CV.CVa` represents the non-dimensional volumetric heat capacities, whereas the control parameter `Energy.k.Ka` represents the vector of thermal conductivities for the different phases (in the listing, phase one is asigned the above values of aluminium).

Bibliography

[1] Alamyane, A. A. and Mohamad, A. A. „Simulation of forced convection in a channel with extended surfaces by the lattice Boltzmann method". In: *Computers & Mathematics with Applications* 59.7 (Apr. 2010), pp. 2421–2430.

[2] Albanakis, C., Missirlis, D., Michailidis, N., Yakinthos, K., Goulas, A., Omar, H., Tsipas, D., and Granier, B. „Experimental analysis of the pressure drop and heat transfer through metal foams used as volumetric receivers under concentrated solar radiation". In: *Experimental Thermal and Fluid Science* 33.2 (Jan. 2009), pp. 246–252.

[3] Almgren, R. F. „Second-Order Phase Field Asymptotics for Unequal Conductivities". In: *SIAM Journal on Applied Mathematics* 59.6 (Jan. 1999), pp. 2086–2107.

[4] Alvarez-Hernandez, A. R. „Combined flow and heat transfer characterization of open cell aluminum foams". Master Thesis. University of Puerto Rico, 2005.

[5] Anderson, D. M. and McFadden, G. B. „Diffuse-Interface methods in Fluid Mechanics". In: *Annual Review of Fluid Mechanics* 30.1 (Jan. 1997), pp. 139–165.

[6] Anderson, J. D. *Computational fluid dynamics*. McGraw-Hill series in aeronautical and aerospace engineering. New York: McGraw-Hill, 2001.

[7] Ashby, M. F. *Metal Foams: A Design Guide*. Boston: Butterworth-Heinemann, 2000.

[8] Baehr, H. D. and Kabelac, S. *Thermodynamik*. Springer-Lehrbuch. Berlin, Heidelberg: Springer Berlin Heidelberg, 2012.

[9] Baehr, H. D. and Stephan, K. *Wärme- und Stoffübertragung*. Springer Lehrbuch. Berlin: Springer Verlag, 2006.

[10] Beckermann, C., Diepers, H. J., Steinbach, I., Karma, A., and Tong, X. „Modeling Melt Convection in Phase-Field Simulations

of Solidification". In: *Journal of Computational Physics* 154 (1999), pp. 468–496.

[11]   Bernoulli, D. *Danielis Bernoulli Hydrodynamica, sive de viribus et motibus fluidorum commentarii : Opus Academicum.* Argentorati: Dulsecker, 1738, p. 304.

[12]   Beugre, D., Calvo, S., Dethier, G., Crine, M., Toye, D., and Marchot, P. „Lattice Boltzmann 3D flow simulations on a metallic foam". In: *Journal of Computational and Applied Mathematics* 234.7 (Aug. 2010), pp. 2128–2134.

[13]   Bhatnagar, P. L., Gross, E. P., and Krook, M. „A Model for Collision Processes in Gases: I. Small Amplitude Process in Charged and Neutral One-Component Systems". In: *Physical Review* 94.3 (1954), pp. 511–525.

[14]   Boettinger, W. J., Warren, J. A., Beckermann, C., and Karma, A. „Phase-Field Simulation of Solidification". In: *Annual Review of Materials Research* 32.1 (2002), pp. 163–194.

[15]   Boltzmann, L. *Vorlesungen über Gastheorie.* Leipzig: Verlag von Johann Ambrosius Barth, 1896.

[16]   Boltzmann, L. „Ueber die sogenannte H-Curve". In: *Mathematische Annalen* 50 (1898), pp. 325 –332.

[17]   Boomsma, K. S. „Metal Foams as Novel Compact High Performance Heat Exchangers for the Cooling of Electronics". PhD Thesis. ETH Zürich, 2002.

[18]   Boomsma, K. S. and Poulikakos, D. „On the effective thermal conductivity of a three-dimensionally structured fluid-saturated metal foam". In: *Heat and Mass Transfer* 44 (2001), pp. 827–836.

[19]   Boomsma, K. S., Poulikakos, D., and Ventikos, Y. „Simulations of flow through open cell metal foams using an idealized periodic cell structure". In: *International Journal of Heat and Fluid Flow* 24.6 (Dec. 2003), pp. 825–834.

[20]   Boomsma, K. S., Poulikakos, D., and Zwick, F. „Metal foams as compact high performance heat exchangers". In: *Mechanics of Materials* 35.12 (Dec. 2003), pp. 1161–1176.

[21]   Bouzidi, M., Firdaouss, M., and Lallemand, P. „Momentum transfer of a Boltzmann-lattice fluid with boundaries". In: *Physics of Fluids* 13.11 (2001), p. 3452.

[22]   Buciuman, F. C. and Kraushaar-Czarnetzki, B. „Ceramic Foam Monoliths as Catalyst Carriers . 1 . Adjustment and Description of the Morphology". In: (2003), pp. 1863–1869.

[23]   Buick, J. M. and Greated, C. A. „Gravity in a lattice Boltzmann model". In: *Physical review. E, Statistical physics, plasmas, fluids, and related interdisciplinary topics* 61.5A (May 2000), pp. 5307–20.

[24]   Burggraf, O. R. „Analytical and numerical studies of the structure of steady separated flows". In: *Journal of Fluid Mechanics* 24.1 (Mar. 1966), pp. 113–151.

[25]   Cahn, J. W. and Hilliard, J. E. „Free Energy of a Nonuniform System. I. Interfacial Free Energy". In: *The Journal of Chemical Physics* 28.2 (1958), p. 258.

[26]   Calmidi, V. V. and Mahajan, R. L. „Forced Convection in High Porosity Metal Foams". In: *Journal of Heat Transfer* 122.3 (2000), p. 557.

[27]   CD-Adapco. *User Guide STAR-CCM+ Version 7.06*. CD-Adapco, 2012.

[28]   Chalkley, D. T. „The Tetrakaidecahedron as the Basisfor the Computation of Cell Volume Density". In: *Science* 13 (1953), pp. 599–600.

[29]   Chapman, S. and Cowling, T. G. *The Mathematical Theory of Nonuniform Gases*. 2nd. New York: Cambridge University Press, 1953.

[30]   Chen, F., Xu, A., Zhang, G., and Li, Y. „Prandtl number effects in MRT Lattice Boltzmann models for shocked and unshocked compressible fluids". In: *Applied Physics* (June 2011), p. 17. arXiv: 1106.0785.

[31]   Chen, F., Flatken, M., Basermann, A., Gerndt, A., Hetheringthon, J., Kruger, T., Matura, G., and Nash, R. W. „Enabling In Situ Pre- and Post-processing for Exascale Hemodynamic Simulations - A Co-design Study with the Sparse Geometry Lattice-Boltzmann Code HemeLB". In: *2012 SC Companion: High Performance Com-*

puting, Networking Storage and Analysis. IEEE, Nov. 2012, pp. 662–
668.

[32]   Chen, L. Q. „Phase-Field Models for Microstructure Evolution".
       In: *Annual Review of Materials Research* 32.1 (2002), pp. 113–140.

[33]   Chen, S. and Doolen, G. D. „Lattice Boltzmann Method for Fluid
       Flows". In: *Annual Review of Fluid Mechanics* 30.1 (Jan. 1998),
       pp. 329–364.

[34]   Chen, S., Martinez, D., and Mei, R. „On boundary conditions in
       lattice Boltzmann methods". In: *Physics of Fluids* 8.9 (1996), p. 2527.

[35]   Chen, S., Chen, H., Martinez, D., and Matthaeus, W. „Lattice
       Boltzmann model for simulation of magnetohydrodynamics". In:
       Physical Review Letters 67.27 (Dec. 1991), pp. 3776–3779.

[36]   Clausius, R. *Über die Art der Bewegung, welche wir Wärme nennen.*
       Poggendorffs, 1857.

[37]   Clausius, R. „Ueber einen auf die Wärme anwendbaren mech-
       anischen Satz". In: *Annalen der Physik und Chemie* 217.9 (1870),
       pp. 124–130.

[38]   Coquard, R. and Baillis, D. „Numerical investigation of conductive
       heat transfer in high-porosity foams". In: *Acta Materialia* 57.18
       (2009), pp. 5466–5479.

[39]   Crouse, B. „Lattice-Boltzmann Strömungssimulationen auf Baum-
       datenstrukturen". Dissertation. TU München, 2003.

[40]   Cruz Ruiz, E. „Modelling of Heat Transfer in Open Cell Metal
       Foams". Master Thesis. University of Puerto Rico, 2004.

[41]   Cueto-Felgueroso, L. and Juanes, R. „Macroscopic Phase-Field
       Model of Partial Wetting: Bubbles in a Capillary Tube". In: *Physical
       Review Letters* 108.14 (Apr. 2012), p. 144502.

[42]   Dai, Z., Nawaz, K., Park, Y. G., Bock, J., and Jacobi, A. M. „Cor-
       recting and extending the Boomsma–Poulikakos effective thermal
       conductivity model for three-dimensional, fluid-saturated metal
       foams". In: *International Communications in Heat and Mass Transfer*
       37.6 (July 2010), pp. 575–580.

[43]   Deville, M. O. and Gatski, T. B. *Mathematical Modeling for Complex
       Fluids and Flows.* Berlin, Heidelberg: Springer Berlin Heidelberg,
       2012.

[44] D'Humières, D. „Generalized lattice-Boltzmann equations". In: *Rarefied Gas Dynamics: Theory and Simulations, Progress in Astronautics and Aeronautics*. Ed. by Shizgal, B D and Weaver, D P. Washington, DC, 1994, pp. 450–458.

[45] D'Humières, D., Ginzburg, I., Krafczyk, M., Lallemand, P., and Luo, L. S. „Multiple-relaxation-time lattice Boltzmann models in three dimensions." In: *Philosophical transactions, Series A, Mathematical, physical and engineering sciences* 360.1792 (Mar. 2002), pp. 437–51.

[46] Diani, A., Bodla, K. K., Rossetto, L., and Garimella, S. V. „Numerical Analysis of Air Flow through Metal Foams". In: *Energy Procedia* 45 (2014), pp. 645–652.

[47] Dietrich, B. „Thermische Charakterisierung von keramischen Schwamm- strukturen für verfahrenstechnische Apparate". Dissertation. Karlsruhe Institute of Technology, 2010.

[48] Dietrich, B., Schell, G., Bucharsky, E. C., Oberacker, R., Hoffmann, M. J., Schabel, W., Kind, M., and Martin, H. „Determination of the thermal properties of ceramic sponges". In: *International Journal of Heat and Mass Transfer* 53.1-3 (Jan. 2010), pp. 198–205.

[49] Donath, S., Feichtinger, Ch., Pohl, Th., Götz, J., and Rüde, U. „A Parallel Free Surface Lattice Boltzmann Method for Large-Scale Applications". In: *Frontiers in Simulation: Simulationstechnique - 18th Symposium (ASIM)*. Ed. by Hülsemann, F., Kowarschik, M., and Rüde, U. Erlangen: SCS Publishing House, 2005, pp. 728–735.

[50] Drew, D. A. „Mathematical Modeling of Two-Phase Flow". In: *Annual Review of Fluid Mechanics* 15.1 (Jan. 1983), pp. 261–291.

[51] Dukhan, N. „Correlations for the pressure drop for flow through metal foam". In: *Experiments in Fluids* (2006), pp. 665–672.

[52] Dukhan, N. and Chen, K. C. „Heat transfer measurements in metal foam subjected to constant heat flux". In: *Experimental Thermal and Fluid Science* 32.2 (Nov. 2007), pp. 624–631.

[53] Dukhan, N. and Patel, K. „Effect of sample's length on flow properties of open-cell metal foam and pressure-drop correlations". In: *Journal of Porous Materials* 18.6 (Oct. 2010), pp. 655–665.

[54] Dukhan, N. and Patel, P. „Equivalent particle diameter and length scale for pressure drop in porous metals". In: *Experimental Thermal and Fluid Science* 32.5 (Apr. 2008), pp. 1059–1067.

[55] Dukhan, N., Quin, P. D., Cruz-Ruiz, E., Scott, E. P., and Ve, M. „One-dimensional heat transfer analysis in open-cell 10-ppi metal foam". In: *Control* 48 (2005), pp. 5112–5120.

[56] Durao, D. F. G., Heitor, M. V., and Pereira, J. C. F. „Measurements of turbulent and periodic flows around a square cross-section cylinder". In: *Experiments in Fluids* 6.5 (1988), pp. 298–304.

[57] Durst, F. *Grundlagen der Strömungsmechanik*. Berlin, Heidelberg: Springer Verlag, 2006.

[58] Echebarria, B., Karma, A., and Plapp, M. „Quantitative phase-field model of alloy solidification". In: *Physical Review E* 70.6 (Dec. 2004), p. 061604.

[59] Edouard, D., Lacroix, M., Huu, C. P., and Luck, F. „Pressure drop modeling on SOLID foam: State-of-the art correlation". In: *Chemical Engineering Journal* 144.2 (Oct. 2008), pp. 299–311.

[60] Einstein, A. „Über die von der molekularkinetischen Theorie der Wärme geforderte Bewegung von in ruhenden Flüssigkeiten suspendierten Teilchen". In: *Annalen der Physik und Chemie* 8 (1905), pp. 549–560.

[61] Erturk, E. „Discussions on driven cavity flow". In: *International Journal for Numerical Methods in Fluids* 60.3 (May 2009), pp. 275–294.

[62] Erturk, E., Corke, T. C., and Gökçöl, C. „Numerical solutions of 2-D steady incompressible driven cavity flow at high Reynolds numbers". In: *International Journal for Numerical Methods in Fluids* 48.7 (July 2005), pp. 747–774.

[63] Ettrich, J., August, A., and Nestler, B. „Open Cell Metal Foams: Measurement and Numerical Modelling of Fluid Flow and Heat Transfer". In: *Cellular Materials: Proceedings CELLMAT 2014, 20-24 Oct. 2014, Dresden, Germany*. Deutsche Gesellschaft für Materialkunde, Fraunhofer-Institut für Angewandte Materialforschung (accepted), 2014.

[64]   Ettrich, J. and Martens, E. „Aufbau eines Prüfstandes zur Vermessung des Wärmeübergangsverhaltens von Metallschaum". In: *Hochschule Karlsruhe – Technik und Wirtschaft, Forschung aktuell* (2012), pp. 25–30.

[65]   Ettrich, J. and Nestler, B. „A Lattice Boltzmann Model combined with a Phase-Field concept for Fluid Dynamics in Complex Cellular Solids". In: *Journal of Computational Physics* submitted (2014).

[66]   Ettrich, J. and Nestler, B. „Diffuse Interface Fluid Flow and Heat Transfer in Cellular Solids". In: *Advanced Engineering Materials* submitted (2014).

[67]   Ettrich, J., Nestler, B., and Rölle, M. „A combined mrt latticeboltzmann and phase-field method for fluid flow and heat transfer simulations in cellular solids". In: *ECCOMAS 2012 - European Congress on Computational Methods in Applied Sciences and Engineering*. 2012, pp. 8461–8472.

[68]   Ettrich, J., Tschukin, O., and Nestler, B. „Diffuse Interface Transient Heat Conduction in Complex Geometries". In: *Modelling and Simulation in Materials Science and Engineering* submitted (2014).

[69]   Ettrich, J., August, A., Rölle, M., and Nestler, B. „Digital Representation of Complex Cellular Structures for Numerical Simulations". In: *Cellular Materials: Proceedings CELLMAT 2014, 20-24 Oct. 2014, Dresden, Germany*. Deutsche Gesellschaft für Materialkunde, Fraunhofer-Institut für Angewandte Materialforschung (accepted), 2014.

[70]   Ettrich, J., Choudhury, A., Tschukin, O., Schoof, E., August, A., and Nestler, B. „Modelling of Transient Heat Conduction with Diffuse Interface Methods". In: *Modelling and Simulation in Materials Science and Engineering* submitted (2014).

[71]   Farhadi, M. and Rahnama, M. „Three-dimensional study of separated flow over a square cylinder by large eddy simulation". In: *Proceedings of the Institution of Mechanical Engineers, Part G: Journal of Aerospace Engineering* 219.3 (Jan. 2005), pp. 225–234.

[72]   Feng, Y. T., Han, K., and Owen, D. R. J. „Coupled lattice Boltzmann method and discrete element modelling of particle transport in turbulent fluid flows: Computational issues". In: *Int. J. Numer. Meth. Engng* 72.June (2007), pp. 1111–1134.

[73]   Ferziger, J. H. and Perić, M. *Computational methods for fluid dynamics*. Berlin: Springer Verlag, 2002.

[74]   Feynman, R. P. *Feynman lectures on physics*. Boston: Addison Wesley Longman, 1970.

[75]   Fourie, J. G. and Du Plessis, J. P. „Pressure drop modelling in cellular metallic foams". In: *Analysis* 57 (2002), pp. 2781 –2789.

[76]   Franke, R. „Numerische Berechnung der instationären Wirbelablösung hinter zylindrischen Körpern". Dissertation. Universität Karlsruhe, 1991.

[77]   Freitas, R. K., Meinke, M., and Schröder, W. „Turbulence Simulation via the Lattice-Boltzmann Method on Hierarchically Refined Meshes". In: *European Conference on Computational Fluid Dynamics ECCOMAS CFD*. Ed. by Wesseling, P, Oñate, E, and Périaux, J. TU Delft, The Netherlands, 2006, pp. 1–12.

[78]   Frisch, U., Hasslacher, B., and Pomeau, Y. „Lattice-Gas Automata for the Navier-Stokes Equation". In: *Physical Review Letters* 56.14 (Apr. 1986), pp. 1505–1508.

[79]   Frisch, U., D'Humières, D., Hasslacher, B., Lallemand, P., Pomeau, Y., and Rivet, J. P. „Lattice Gas Hydrodynamics in Two a nd Three Dimensions". In: *Complex Systems* 1 (1987), pp. 649–707.

[80]   Frohn, A. *Einführung in die kinetische Gastheorie*. Wiesbaden: Akademische Verlagsgesellschaft, 1979.

[81]   Gallivan, M. A., Noble, D. R., Georgiadis, J. G., and Buckius, R. O. „An evaluation of the bounce-back boundary condition for lattice boltzmann simulations". In: *International Journal* 25.June 1996 (1997), pp. 249–263.

[82]   Garrido, I. G., Patcas, F. C., Lang, S., and Kraushaar-Czarnetzki, B. „Mass transfer and pressure drop in ceramic foams: A description for different pore sizes and porosities". In: *Chemical Engineering Science* 63.21 (Nov. 2008), pp. 5202–5217.

[83]   Gassner, G., Lörcher, F., and Munz, C. D. „A contribution to the construction of diffusion fluxes for finite volume and discontinuous Galerkin schemes". In: *Journal of Computational Physics* 224.2 (June 2007), pp. 1049–1063.

[84] Ghia, U., Ghia, K. N., and Shin, C. T. „High-Re Solutions for Incompressible Flow Using the Navier-Stokes Equations Multigrid Method". In: *Journal of Computational Physics* 48 (1982), pp. 387–411.

[85] Ghosh, I. „Heat transfer correlation for high-porosity open-cell foam". In: *International Journal of Heat and Mass Transfer* 52.5-6 (Feb. 2009), pp. 1488–1494.

[86] Gibson, L. J. and Ashby, M. F. *Cellular solids*. 2. Cambridge solid state science series. Cambridge: Cambridge Univ. Press, 2001.

[87] Ginzburg, I. and D'Humières, D. „Multireflection boundary conditions for lattice Boltzmann models". In: *Physical Review E* 68.6 (Dec. 2003), pp. 1–30.

[88] Girlich, D., Hackeschmidt, K., and Kühn, C. „Bestimmung des konvektiven Wärmeübergangs offenporiger Metallschäume". In: *Konstruktion* Heft 1/2 (2005), pp. 1–10.

[89] Girlich, D., Kühn, C., and Hackeschmidt, K. „Berechnung des Druckverlustes für offenporigen Metallschäume". In: *Zeitschrift Konstruktion* Heft 4 (2004).

[90] Gräf, I., Rühl, A.-K., and Kraushaar-Czarnetzki, B. „Experimental study of heat transport in catalytic sponge packings by monitoring spatial temperature profiles in a cooled-wall reactor". In: *Chemical Engineering Journal* 244 (May 2014), pp. 234–242.

[91] Guo, Z., Zheng, C., and Shi, B. „An extrapolation method for boundary conditions in lattice Boltzmann method". In: *Physics of Fluids* 14.6 (2002), p. 2007.

[92] Guo, Z., Zheng, C., and Shi, B. „Discrete lattice effects on the forcing term in the lattice Boltzmann method". In: *Physical Review E* 65.4 (Apr. 2002), p. 046308.

[93] Gurtin, M. E. and Voorhees, P. W. „The thermodynamics of evolving interfaces far from equilibrium". In: *Acta Materialia* 44.1 (Jan. 1996), pp. 235–247.

[94] Han, X. H., Wang, Q., Park, Y. G., T'Joen, Ch., Sommers, A., and Jacobi, A. „A Review of Metal Foam and Metal Matrix Composites for Heat Exchangers and Heat Sinks". In: *Heat Transfer Engineering* 33.12 (Sept. 2012), pp. 991–1009.

[95] Hänel, D. *Molekulare Gasdynamik*. Berlin, Heidelberg: Springer Verlag, 2004.

[96] Hardy, J., Pazzis, O. de, and Pomeau, Y. „Molecular dynamics of a classical lattice gas: Transport properties and time correlation functions". In: *Physical Review A* 13.5 (May 1976), pp. 1949–1961.

[97] Harting, J., Chin, J., Venturoli, M., and Coveney, P. V. „Large-scale lattice Boltzmann simulations of complex fluids: advances through the advent of computational Grids." In: *Philosophical transactions. Series A, Mathematical, physical, and engineering sciences* 363.1833 (Aug. 2005), pp. 1895–915.

[98] Hecht, M. and Harting, J. „Implementation of on-site velocity boundary conditions for D3Q19 lattice Boltzmann simulations". In: *Journal of Statistical Mechanics: Theory and Experiment* 2010.01 (Jan. 2010), P01018.

[99] Heuveline, V., Krause, M. J., and Latt, J. „Towards a hybrid parallelization of lattice Boltzmann methods". In: *Computers & Mathematics with Applications* 58.5 (Sept. 2009), pp. 1071–1080.

[100] Hipke, T., Lange, G., and Poss, R. *Taschenbuch für Aluminiumschäume*. Düsseldorf: Alu-Media, 2007.

[101] Hirsch, C. *Numerical Computation of Internal and External Flows. Fundamentals of Computational Fluid Dynamics*. 2. ed. Butterworth-Heinemann, 2007.

[102] Hoffmann, K. A. and Chiang, S. T. *Computational fluid dynamics - Volume I*. Engineering Education System, 2000.

[103] Hötzer, J., Jainta, M., Vondrous, A., Ettrich, J., August, A., Stubenvoll, D., Reichardt, M., Selzer, M., and Nestler, B. „Phase-field simulations of large-scale microstructures by integrated parallel algorithms". In: *High Performance Computing in Science and Engineering '14, Transactions of the High Performance Computing Center, Stuttgart (HLRS) 2014*. Ed. by Nagel, W. E., Kröner, D. H., and Resch, M. M. (in print). Heidelberg Dordrecht London New York: Springer, 2014.

[104] Hou, S., Sterling, J., Chen, S., and Doolen, G. D. „A Lattice Boltzmann Subgrid Model for High Reynolds Number Flows". In: (Jan. 1994), pp. 1–18. arXiv: 9401004 [comp-gas].

[105] Huu, T. T., Lacroix, M., Huu, C. P., Schweich, D., and Edouard, D. „Towards a more realistic modeling of solid foam Use of the pentagonal dodecahedron geometry". In: *Chemical Engineering Science* 64.24 (2009), pp. 5131–5142.

[106] Inamuro, T., Yoshino, M., and Ogino, F. „Accuracy of the lattice Boltzmann method for small Knudsen number with finite Reynolds number". In: *Physics of Fluids* 9.11 (1997), p. 3535.

[107] Inayat, A., Freund, H. J., Zeiser, Th., and Schwieger, W. „Determining the specific surface area of ceramic foams: The tetrakaidecahedra model revisited". In: *Chemical Engineering Science* 66.6 (Mar. 2011), pp. 1179–1188.

[108] Jami, M., Mezrhab, A., Bouzidi, M., and Lallemand, P. „Lattice Boltzmann method applied to the laminar natural convection in an enclosure with a heat-generating cylinder conducting body". In: *International Journal of Thermal Sciences* 46.1 (Jan. 2007), pp. 38–47.

[109] Jun, C. and Xiu-Lan, H. „A Lattice Boltzmann Model for Fluid-Solid Coupling Heat Transfer in Fractal Porous Media". In: *Chinese Physics Letters* 26.6 (June 2009), p. 064401.

[110] Kabelac, S. *VDI-Wärmeatlas*. VDI Buch. Berlin, Heidelberg, New-York: Springer Verlag, 2006.

[111] Karma, A. „Phase-Field Formulation for Quantitative Modeling of Alloy Solidification". In: *Physical Review Letters* 87.11 (Aug. 2001), p. 115701.

[112] Karma, A. and Rappel, W. J. „Phase-field method for computationally efficient modeling of solidification with arbitrary interface kinetics". In: *Physical Review E* 53.4 (Apr. 1996), R3017–R3020.

[113] Kaviany, M. *Principles of heat transfer in porous media*. Mechanical engineering series. New York: Springer, 1995.

[114] Kopanidis, A., Theodorakakos, A., Gavaises, E., and Bouris, D. „3D numerical simulation of flow and conjugate heat transfer through a pore scale model of high porosity open cell metal foam". In: *International Journal of Heat and Mass Transfer* 53.11-12 (May 2010), pp. 2539–2550.

[115] Krafczyk, M. „Gitter-Boltzmann-Methoden: Von der Theorie zur Anwendung". Habilitationsschrift. TU München, 2001.

[116] Kroeger, P. „Implementierung und Validierung eines adaptiven Gitterverfahrens für einen finiten Differenzen-Solver". Bachelor Thesis. Hochschule Karlsruhe, 2012.

[117] Kupershtokh, A. „New method of incorporating a body force term into lattice Boltzmann equation". In: *Proceedings of the 5th International EHD Workshop5th International EHD Workshop*. Poitiers, France: University of Poitiers, 2004, pp. 241–246.

[118] Lallemand, P. and Luo, L. S. „Theory of the lattice Boltzmann method: Dispersion, dissipation, isotropy, Galilean invariance, and stability". In: *Physical Review E* 61.6 (2000), pp. 6546–6562.

[119] Lallemand, P. and Luo, L. S: „Hybrid finite-difference thermal lattice boltzmann equation". In: *International Journal of Modern Physics B* 17.1 & 2 (2003), pp. 41–47.

[120] Lallemand, P. and Luo, L. S. „Theory of the lattice Boltzmann method: Acoustic and thermal properties in two and three dimensions". In: *Physical Review E* 68.3 (Sept. 2003).

[121] Lätt, J. „Hydrodynamic limit of lattice Boltzmann equations". PhD Thesis. Université de Genève, 2007.

[122] Lienhard IV, J. H. and Lienhard V, J. H. *A heat transfer textbook.* Cambridge, Mass: Phlogiston Press, 2003.

[123] Liou, M. F. and Greber, I. „Mesh-Based Microstructure Representation Algorithm for Simulating Pore-scale Transport Phenomena in Porous Media". In: *Methodology* (2009).

[124] Liu, T., Liu, G., Ge, Y., Wu, H., and Wu, W. „Extended lattice boltzmann equation for simulation of flows around bluff bodies in high reynolds number". In: *BBAA VI International Colloquium on: Bluff Body Aerodynamics & Applications, Milano, Italy, July 20-24 2008*. Dvm. Milano, 2008.

[125] Lu, T. J., Stone, H. A., and Ashby, M. F. „Heat transfer in open-cell metal foams". In: *Acta Materialia* 46.10 (June 1998), pp. 3619–3635.

[126] Lu, W., Zhao, C., and Tassou, S. „Thermal analysis on metal-foam filled heat exchangers. Part I: Metal-foam filled pipes". In:

International Journal of Heat and Mass Transfer 49.15-16 (July 2006), pp. 2751–2761.

[127] Luo, L. S. „Unified Theory of Lattice Boltzmann Models for Non-ideal Gases". In: *Physical Review Letters* 81.8 (Aug. 1998), pp. 1618–1621.

[128] Luo, L. S: „The Lattice-Gas and Lattice-Boltzmann Methods: Past, Present and Future". In: *Proceedings of the "International Conference on Applied Computational Fluid Dynamics", Beijing, China, October 17-20.* Ed. by Wu, J H and Zhu, Z J. 2000, pp. 52–83.

[129] Luo, L. S: *Theory of the lattice Boltzmann Equation.* Tech. rep. Beijing: China Center of Advanced Science and technology, 2000.

[130] Lyn, D. A., Einav, S., Rodi, W., and Park, J. H. „A laser-Doppler velocimetry study of ensemble-averaged characteristics of the turbulent near wake of a square cylinder". In: *Journal of Fluid Mech.* 304 (1995), pp. 285–319.

[131] Maier, R. S., Bernard, R. S., and Grunau, D. W. „Boundary conditions for the lattice Boltzmann method". In: *Physics of Fluids* 8.7 (1996), p. 1788.

[132] Mancin, S., Zilio, C., Cavallini, A., and Rossetto, L. „Heat transfer during air flow in aluminum foams". In: *International Journal of Heat and Mass Transfer* 53.21-22 (Oct. 2010), pp. 4976–4984.

[133] Mancin, S., Zilio, C., Cavallini, A., and Rossetto, L. „Pressure drop during air flow in aluminum foams". In: *International Journal of Heat and Mass Transfer* 53.15-16 (July 2010), pp. 3121–3130.

[134] Mancin, S., Zilio, C., Diani, A., and Rossetto, L. „Experimental air heat transfer and pressure drop through copper foams". In: *Experimental Thermal and Fluid Science* 36 (Jan. 2012), pp. 224–232.

[135] Mancin, S., Zilio, C., Diani, A., and Rossetto, L. „Air forced convection through metal foams: Experimental results and modeling". In: *International Journal of Heat and Mass Transfer* 62 (July 2013), pp. 112–123.

[136] Martys, N., Shan, X., and Chen, H. „Evaluation of the external force term in the discrete Boltzmann equation". In: *Physical Review E* 58.5 (Nov. 1998), pp. 6855–6857.

[137] Massaioli, F., Benzi, R., and Succi, S. „Exponential Tails in Two-Dimensional Rayleigh-Bénard Convection". In: *1Europhysics Letters (EPL)* 21.3 (1993), pp. 305–310.

[138] Mavriplis, D. J. „Multigrid solution of the steady-state lattice Boltzmann equation". In: *Computers & Fluids* 35.8-9 (Sept. 2006), pp. 793–804.

[139] Maxwell, J. C. „On the Dynamical Theory of Gases". In: *Philosophical Transactions of the Royal Society of London* 157.1867 (1867), pp. 49–88.

[140] McFadden, G. B., Wheeler, A. A., and Anderson, D. M. „Thin interface asymptotics for an energy/entropy approach to phase-field models with unequal conductivities". In: *Physica D: Nonlinear Phenomena* 144.1-2 (Sept. 2000), pp. 154–168.

[141] McNamara, G. R. and Zanetti, G. „Use of the Boltzmann Equation to Simulate Lattice-Gas Automata". In: *Physical Review Letters* 61.20 (Nov. 1988), pp. 2332–2335.

[142] Medvedev, D. and Kassner, K. „Lattice Boltzmann scheme for crystal growth in external flows". In: *Physical Review E* 72.5 (Nov. 2005), p. 056703.

[143] Medvedev, D. and Kassner, K. „Lattice-Boltzmann scheme for dendritic growth in presence of convection". In: *Journal of Crystal Growth* 275.1-2 (Feb. 2005), e1495–e1500.

[144] Medvedev, D., Varnik, F., and Steinbach, I. „Simulating Mobile Dendrites in a Flow". In: *Procedia Computer Science* 18 (Jan. 2013), pp. 2512–2520.

[145] Mennerich, C. „Phase-field modeling of multi-domain evolution in ferromagnetic shape memory alloys and of polycrystalline thin film growth". PhD Thesis. Karlsruhe Institute of Technology, 2013.

[146] Mezrhab, A., Bouzidi, M., and Lallemand, P. „Hybrid lattice Boltzmann finite difference simulation of convective flows". In: *Computers & Fluids* 33.4 (2004), pp. 623–641.

[147] Mezrhab, A., Amine Moussaoui, M., Jami, M., Naji, H., and Bouzidi, M. „Double MRT thermal lattice Boltzmann method for simulating convective flows". In: *Physics Letters A* 374.34 (July 2010), pp. 3499–3507.

[148] Miller, W., Rasin, I., and Succi, S. „Lattice Boltzmann phase-field modelling of binary-alloy solidification". In: *Physica A: Statistical Mechanics and its Applications* 362.1 (Mar. 2005), pp. 78–83.

[149] Miwa, S. and Revankar, S. T. „Hydrodynamic Characterization of Nickel Metal Foam, Part 1: Single-Phase Permeability". In: *Transport in Porous Media* 80.2 (Mar. 2009), pp. 269–279.

[150] Mohamad, A. A. and Kuzmin, A. „A critical evaluation of force term in lattice Boltzmann method, natural convection problem". In: *International Journal of Heat and Mass Transfer* 53.5-6 (Feb. 2010), pp. 990–996.

[151] Mohamad, A. A. and Succi, S. „A note on equilibrium boundary conditions in lattice Boltzmann fluid dynamic simulations". In: *The European Physical Journal Special Topics* 171.1 (May 2009), pp. 213–221.

[152] Mohammadi Pirouz, M., Farhadi, M., Sedighi, K., Nemati, H., and Fattahi, E. „Lattice Boltzmann simulation of conjugate heat transfer in a rectangular channel with wall-mounted obstacles". In: *Scientia Iranica* 18.2 (Apr. 2011), pp. 213–221.

[153] Moussaoui, M. A., Jami, M., Mezrhab, A., and Naji, H. „MRT-Lattice Boltzmann simulation of forced convection in a plane channel with an inclined square cylinder". In: *International Journal of Thermal Sciences* 49.1 (Jan. 2010), pp. 131–142.

[154] Nestler, B. „A 3D parallel simulator for crystal growth and solidification in complex alloy systems". In: *Journal of Crystal Growth* 275.1-2 (Feb. 2005), e273–e278.

[155] Nestler, B., Aksi, A., and Selzer, M. „Combined Lattice Boltzmann and phase-field simulations for incompressible fluid flow in porous media". In: *Mathematics and Computers in Simulation* 80.7 (Mar. 2010), pp. 1458–1468.

[156] Nestler, B., Garcke, H., and Stinner, B. „Multicomponent alloy solidification: Phase-field modeling and simulations". In: *Physical Review E* 71.4 (Apr. 2005), p. 041609.

[157] Ni, J. and Beckermann, C. „A volume-averaged two-phase model for transport phenomena during solidification". In: *Metallurgical and Materials Transactions B* (1991).

[158]   Nicoli, M., Plapp, M., and Henry, H. „Tensorial mobilities for accurate solution of transport problems in models with diffuse interfaces". In: *Physical Review E* 84.4 (Oct. 2011), pp. 1–6.

[159]   Nitsche, W. and Brunn, A. *Strömungsmesstechnik*. Berlin, Heidelberg: Springer Verlag, 2006.

[160]   Noble, D. R., Chen, S., Georgiadis, J. G., and Buckius, R. O. „A consistent hydrodynamic boundary condition for the lattice Boltzmann method". In: *Physics of Fluids* 7.1 (1995), p. 203.

[161]   Nourgaliev, R., Dinh, T. N., Theofanous, T. G., and Joseph, D. „The lattice Boltzmann equation method: theoretical interpretation, numerics and implications". In: *International Journal of Multiphase Flow* 29.1 (Jan. 2003), pp. 117–169.

[162]   Oertel, H., Böhle, M., and Dohrmann, U. *Strömungsmechanik*. Wiesbaden: Vieweg, 2009.

[163]   Ohno, M. and Matsuura, K. „Quantitative phase-field modeling for dilute alloy solidification involving diffusion in the solid". In: *Physical Review E* 79.3 (Mar. 2009), p. 031603.

[164]   Onsager, L. „Reciprocal Relations in Irreversible Processes. I." In: *Physical Review* 37.4 (Feb. 1931), pp. 405–426.

[165]   Onsager, L. „Reciprocal Relations in Irreversible Processes. II." In: *Physical Review* 38.12 (Dec. 1931), pp. 2265–2279.

[166]   Onstad, A. J., Elkins, Ch. J., Medina, F., Wicker, R. B., and Eaton, J. K. „Full-field measurements of flow through a scaled metal foam replica". In: *Experiments in Fluids* 50.6 (Dec. 2010), pp. 1571–1585.

[167]   Ozisik, M. N. *Heat Conduction*. 2nd. New York: John Wiley & Sons, Inc., 1993.

[168]   Patankar, S. V. *Numerical heat transfer and fluid flow*. Series in computational methods in mechanics and thermal sciences. New York: Hemisphere Publ. Co., 1980.

[169]   Plapp, M. „Remarks on some open problems in phase-field modelling of solidification". In: (Apr. 2010), p. 22. arXiv: 1004.4502.

[170]   Polifke, W. and Kopitz, J. *Wärmeübertragung: Grundlagen, analytische und numerische Methoden*. München: Pearson Deutschland, 2009, p. 609.

[171]   Pope, SB. *Turbulent flows*. New York: Cambridge University Press, 2000.

[172]   Premnath, K. N., Pattison, M. J., and Banerjee, S. „Dynamic subgrid scale modeling of turbulent flows using lattice-Boltzmann method". In: *Physica A: Statistical Mechanics and its Applications* 388.13 (July 2009), pp. 2640–2658.

[173]   Press, W. H., Teukolsky, S. A., Vetterling, W. T., and Flannery, B. P. *Numerical recipes*. 3. ed. Cambridge: Cambridge Univ. Press, 2007.

[174]   Qian, Y. H., D'Humières, D., and Lallemand, P. „Lattice BGK Models for Navier-Stokes Equation". In: *Europhysics Letters (EPL)* 17.6 (Feb. 1992), pp. 479–484.

[175]   Rasin, I., Miller, W., and Succi, S. „Phase-field lattice kinetic scheme for the numerical simulation of dendritic growth". In: *Physical Review E* (2005), pp. 1–10.

[176]   Rasin, I., Succi, S., and Miller, W. „A multi-relaxation lattice kinetic method for passive scalar diffusion". In: *Journal of Computational Physics* 206.2 (July 2005), pp. 453–462.

[177]   Rhode, M. „Extending the lattice-Boltzmann method: Novel techniques for local grid refinement and boundary conditions". PhD thesis. 2004.

[178]   Rölle, M. „Füllalgorithmen zur Generierung schaumartiger Strukturen in 3D". Bachelor Thesis. Hochschule Karlsruhe - Technik und Wirtschaft, 2010.

[179]   Schlegel, A., Benz, P., and Buser, S. „Wärmeübertragung und Druckabfall in keramischen Schaumstrukturen bei erzwungener Strömung". In: *Wärme- und Stoffübertragung* 28 (1993), pp. 259–266.

[180]   Schlichting, H., Gersten, K., Krause, E., and Oertel, H. *Grenzschicht-Theorie*. Springer Verlag, 2006.

[181]   Schlünder, E. U. and Martin, H. *Einführung in die Wärmeübertragung*. 8. Vieweg, 1995, p. 242.

[182]   Schneiderer, S. „Effiziente parallele Lattice-Boltzmann Simulation für turbulente Strömungen". Diplomarbeit. Universität Stuttgart, 2006.

[183]   Sertkaya, A. A., Altınısık, K., and Dincer, K. „Experimental investigation of thermal performance of aluminum finned heat

exchangers and open-cell aluminum foam heat exchangers". In: *Experimental Thermal and Fluid Science* 36 (Jan. 2012), pp. 86–92.

[184] Shan, Xiaowen. „Simulation of Rayleigh-Bénard convection using a lattice Boltzmann method". In: *Physical Review E* 55.3 (1997), pp. 2780–2788.

[185] Singer-Loginova, I. and Singer, H. M. „The phase field technique for modeling multiphase materials". In: *Reports on Progress in Physics* 71.10 (2008), p. 106501.

[186] Skordos, P. „Initial and boundary conditions for the lattice Boltzmann method". In: *Physical Review E* 48.6 (Dec. 1993), pp. 4823–4842.

[187] Smagorinsky, J. „Monthly weather". In: *Monthly Weather Review* 91.3 (1963), pp. 99–164.

[188] Smoluchowski, M. „Zur kinetischen Theorie der brownschen Molekularbewegung und der Suspensionen". In: *Annalen der Physik* 326.14 (1906), pp. 756–780.

[189] Spanos, T . J . T . *The Thermophysics of Porous Media*. Chapman & Hall/CRC, 2002.

[190] Srinivas, Y., Biswas, G., Parihar, A. S., and Ranjan, R. „Large-Eddy Simulation of High Reynolds Number Turbulent Flow Past a Square Cylinder". In: *Journal of Engineering Mechanics* 132.3 (Mar. 2006), pp. 327–335.

[191] Steinbach, I., Pezzolla, F., and Nestler, B. „A phase field concept for multiphase systems". In: *Physica D: Nonlinear ...* 94 (1996), pp. 135–147.

[192] Stephens, M. A. „EDF statistics for goodness of fit and some comparisons". In: *Journal of the American statistical Association* 69.347 (1974), pp. 730–737.

[193] Strohm, K. „Modifizierte Navier–Stokes–Gleichungen für Phasenfeldmodelle". Diplomarbeit. Karlsruher Institut für Technologie, 2012.

[194] Succi, S. *The Lattice Boltzmann Equation*. Oxford science publications. Oxford: Clarendon Press, 2001.

[195] Succi, S., Bella, G., and Papetti, F. „Lattice Kinetic Theory for Numerical Combustion". In: *Journal of Scientific Computing* 12.4 (1997), pp. 395–408.

[196] Sukop, M. C. and Thorne Jr., D. T. *Lattice Boltzmann Modeling: An Introduction for Geoscientists and Engineers*. 2nd. Berlin, Heidelberg: Springer Verlag, 2006, p. 172.

[197] Sullivan, R., Ghosn, L., and Lerch, B. „A general tetrakaidecahedron model for open-celled foams". In: *International Journal of Solids and Structures* 45.6 (Mar. 2008).

[198] Sun, Ch. „Lattice-Boltzmann models for high speed flows". In: *Physical Review E* 58.6 (Dec. 1998), pp. 7283–7287.

[199] Tadrist, L. „About the use of fibrous materials in compact heat exchangers". In: *Experimental Thermal and Fluid Science* 28.2-3 (2004), pp. 193–199.

[200] Tannehill, J. C., Anderson, D. A., and Pletcher, R. H. *Computational fluid mechanics and heat transfer*. 2nd. Taylor & Francis, 1997.

[201] Tarokh, A. A., Mohamad, A. A., and Jiang, L. „Simulation of Conjugate Heat Transfer Using the Lattice Boltzmann Method". In: *Numerical Heat Transfer, Part A: Applications* 63.3 (Jan. 2013), pp. 159–178.

[202] Thümmes, O. „Experimentelle Untersuchungen der fluidmechanischen und thermischen Eigenschaften von Metallschäumen". Bachelor Thesis. Karlsruhe University of Applied Siences, 2011.

[203] Tölke, J., Freudiger, S., and Krafczyk, M. „An adaptive scheme using hierarchical grids for lattice Boltzmann multi-phase flow simulations". In: *Computers and Fluids* 35.8-9 (Sept. 2006), pp. 820–830.

[204] Tölke, J., Krafczyk, M., and Rank, E. „A Multigrid-Solver for the Discrete Boltzmann Equation". In: *Journal of Statistical Physics* 107.April (2002).

[205] Tong, X., Beckermann, C., Karma, A., and Li, Q. „Phase-field simulations of dendritic crystal growth in a forced flow". In: *Physical Review E* 63.6 (May 2001), pp. 1–16.

[206] Treeck, C. A. van. „Gebäudemodell-basierte Simulation von Raum-luftströmungen". PhD Thesis. Technische Universität München, 2004.

[207] Treeck, C. A. van, Rank, E., Krafczyk, M., Tölke, J., and Nachtwey, B. „Extension of a hybrid thermal LBE scheme for large-eddy simulations of turbulent convective flows". In: *Computers & Fluids* 35.8-9 (Sept. 2006), pp. 863–871.

[208] Vengadesan, S. and Nakayama, A. „Evaluation of LES models for flow over bluff body from engineering application perspective". In: *Sadhana* 30.1 (Feb. 2005), pp. 11–20.

[209] Viggen, E. M. „The Lattice Boltzmann Method with Applications in Acoustics". Master Thesis. Norwegian University of Science and Technology, 2009.

[210] Viggen, E. M. „The lattice Boltzmann method: Fundamentals and acoustics". PhD Thesis. Norwegian University of Science and Technology, 2014.

[211] Vondrous, A., Selzer, M., Hötzer, J., and Nestler, B. „Parallel computing for phase-field models". In: *International Journal of High Performance Computing Applications* (June 2013).

[212] Wang, J., Wang, M., and Li, Z. „A lattice Boltzmann algorithm for fluid–solid conjugate heat transfer". In: *International Journal of Thermal Sciences* 46.3 (Mar. 2007), pp. 228–234.

[213] Wellein, G., Lammers, P., Hager, G., Donath, S., and Zeiser, T. „Towards optimal performance for lattice Boltzmann applications on terascale computers". In: *Parallel Computational Fluid Dynamics 2005: Theory and Applications*. Ed. by Deane, A., Brenner, G., Emerson, D., McDonough, J., Tromeur-Dervout, D., Satofuka, N., Ecer, A., and Periaux, J. Elsevier B.V., 2006, pp. 31–40.

[214] Wilcox, D. C. *Turbulence modeling for CFD*. La Cañada, California: DCW Industries, 1994.

[215] Wolf-Gladrow, D. A. *Lattice-Gas Cellular Automata and Lattice Boltz-mann Models - An Introduction*. Springer Verlag, 2005, p. 311.

[216] Worthing, R. A., Mozer, J., and Seeley, G. „Stability of lattice Boltz-mann methods in hydrodynamic regimes". In: *Physical Review E* 56.2 (1997), pp. 2243–2253.

[217] Yarin, L.P. *The Pi-Theorem*. Berlin, Heidelberg: Springer Berlin Heidelberg, 2012.

[218] Younis, L.B. and Viskanta, R. „Experimental determination of the volumetric heat transfer coefficient between stream of air and ceramic foam". In: *International Journal of Heat and Mass Transfer* 36.6 (Jan. 1993), pp. 1425–1434.

[219] Yu, D., Mei, R., Luo, L. S., and Shyy, W. „Viscous flow computations with the method of lattice Boltzmann equation". In: *Progress in Aerospace Sciences* 39.5 (July 2003), pp. 329–367.

[220] Yu, H. „Lattice Boltzmann Equation Simulations of Turbulence, Mixing, and Combustion". PhD thesis. Texas A&M University, 2004.

[221] Yu, Z. and Fan, L. S. „An interaction potential based lattice Boltzmann method with adaptive mesh refinement (AMR) for two-phase flow simulation". In: *Journal of Computational Physics* 228.17 (Sept. 2009), pp. 6456–6478.

[222] Zaragoza, G. and Goodall, R. „Metal Foams with Graded Pore Size for Heat Transfer Applications". In: *Advanced Engineering Materials* 15.3 (Mar. 2013), pp. 123–128.

[223] Zhao, C., Lu, W., and Tassou, S. „Thermal analysis on metal-foam filled heat exchangers. Part II: Tube heat exchangers". In: *International Journal of Heat and Mass Transfer* 49.15-16 (July 2006), pp. 2762–2770.

[224] Zou, Q. and He, X. „On pressure and velocity boundary conditions for the lattice Boltzmann BGK model". In: *Physics of Fluids* 9.6 (1997), pp. 1591–1598.

Schriftenreihe
des Instituts für Angewandte Materialien

ISSN 2192-9963

Die Bände sind unter www.ksp.kit.edu als PDF frei verfügbar
oder als Druckausgabe bestellbar.

Band 1   Prachai Norajitra
Divertor Development for a Future Fusion Power Plant. 2011
ISBN 978-3-86644-738-7

Band 2   Jürgen Prokop
**Entwicklung von Spritzgießsonderverfahren zur Herstellung
von Mikrobauteilen durch galvanische Replikation.** 2011
ISBN 978-3-86644-755-4

Band 3   Theo Fett
**New contributions to R-curves and bridging stresses –
Applications of weight functions.** 2012
ISBN 978-3-86644-836-0

Band 4   Jérôme Acker
**Einfluss des Alkali/Niob-Verhältnisses und der Kupferdotierung
auf das Sinterverhalten, die Strukturbildung und die Mikro-
struktur von bleifreier Piezokeramik $(K_{0,5}Na_{0,5})NbO_3$.** 2012
ISBN 978-3-86644-867-4

Band 5   Holger Schwaab
**Nichtlineare Modellierung von Ferroelektrika unter
Berücksichtigung der elektrischen Leitfähigkeit.** 2012
ISBN 978-3-86644-869-8

Band 6   Christian Dethloff
**Modeling of Helium Bubble Nucleation and Growth
in Neutron Irradiated RAFM Steels.** 2012
ISBN 978-3-86644-901-5

Band 7   Jens Reiser
**Duktilisierung von Wolfram. Synthese, Analyse und
Charakterisierung von Wolframlaminaten aus Wolframfolie.** 2012
ISBN 978-3-86644-902-2

Band 8   Andreas Sedlmayr
**Experimental Investigations of Deformation Pathways
in Nanowires.** 2012
ISBN 978-3-86644-905-3

Band 9   Matthias Friedrich Funk
Microstructural stability of nanostructured fcc metals during cyclic deformation and fatigue. 2012
ISBN 978-3-86644-918-3

Band 10   Maximilian Schwenk
Entwicklung und Validierung eines numerischen Simulationsmodells zur Beschreibung der induktiven Ein- und Zweifrequenzrandschichthärtung am Beispiel von vergütetem 42CrMo4. 2012
ISBN 978-3-86644-929-9

Band 11   Matthias Merzkirch
Verformungs- und Schädigungsverhalten der verbundstranggepressten, federstahldrahtverstärkten Aluminiumlegierung EN AW-6082. 2012
ISBN 978-3-86644-933-6

Band 12   Thilo Hammers
Wärmebehandlung und Recken von verbundstranggepressten Luftfahrtprofilen. 2013
ISBN 978-3-86644-947-3

Band 13   Jochen Lohmiller
Investigation of deformation mechanisms in nanocrystalline metals and alloys by in situ synchrotron X-ray diffraction. 2013
ISBN 978-3-86644-962-6

Band 14   Simone Schreijäg
Microstructure and Mechanical Behavior of Deep Drawing DC04 Steel at Different Length Scales. 2013
ISBN 978-3-86644-967-1

Band 15   Zhiming Chen
Modelling the plastic deformation of iron. 2013
ISBN 978-3-86644-968-8

Band 16   Abdullah Fatih Çetinel
Oberflächendefektausheilung und Festigkeitssteigerung von niederdruckspritzgegossenen Mikrobiegebalken aus Zirkoniumdioxid. 2013
ISBN 978-3-86644-976-3

Band 17   Thomas Weber
Entwicklung und Optimierung von gradierten Wolfram/ EUROFER97-Verbindungen für Divertorkomponenten. 2013
ISBN 978-3-86644-993-0

Band 18   Melanie Senn
Optimale Prozessführung mit merkmalsbasierter Zustandsverfolgung. 2013
ISBN 978-3-7315-0004-9

Band 29   Gunnar Picht
           **Einfluss der Korngröße auf ferroelektrische Eigenschaften
           dotierter Pb(Zr$_{1-x}$Ti$_x$)O$_3$ Materialien.** 2013
           ISBN 978-3-7315-0106-0

Band 30   Esther Held
           **Eigenspannungsanalyse an Schichtverbunden
           mittels inkrementeller Bohrlochmethode.** 2013
           ISBN 978-3-7315-0127-5

Band 31   Pei He
           **On the structure-property correlation and the evolution
           of Nanofeatures in 12-13.5% Cr oxide dispersion strengthened
           ferritic steels.** 2014
           ISBN 978-3-7315-0141-1

Band 32   Jan Hoffmann
           **Ferritische ODS-Stähle – Herstellung,
           Umformung und Strukturanalyse.** 2014
           ISBN 978-3-7315-0157-2

Band 33   Wiebke Sittel
           **Entwicklung und Optimierung des Diffusionsschweißens
           von ODS Legierungen.** 2014
           ISBN 978-3-7315-0182-4

Band 34   Osama Khalil
           **Isothermes Kurzzeitermüdungsverhalten der hoch-warmfesten
           Aluminium-Knetlegierung 2618A (AlCu2Mg1,5Ni).** 2014
           ISBN 978-3-7315-0208-1

Band 35   Magalie Huttin
           **Phase-field modeling of the influence of mechanical stresses on
           charging and discharging processes in lithium ion batteries.** 2014
           ISBN 978-3-7315-0213-5

Band 36   Christoph Hage
           Grundlegende Aspekte des 2K-Metallpulverspritzgießens. 2014
           ISBN 978-3-7315-0217-3

Band 37   Bartłomiej Albiński
           **Instrumentierte Eindringprüfung bei Hochtemperatur
           für die Charakterisierung bestrahlter Materialien.** 2014
           ISBN 978-3-7315-0221-0

Band 38   Tim Feser
           **Untersuchungen zum Einlaufverhalten binärer alpha-
           Messinglegierungen unter Ölschmierung in Abhängigkeit
           des Zinkgehaltes.** 2014
           ISBN 978-3-7315-0224-1

Band 39   Jörg Ettrich
Fluid Flow and Heat Transfer in Cellular Solids. 2014
ISBN 978-3-7315-0241-8